ASSESSING REQUIREMENTS FOR PEACEKEEPING, HUMANITARIAN ASSISTANCE, AND DISASTER RELIEF

BRUCE R. PIRNIE

CORAZON M. FRANCISCO

Prepared for the
Office of the Secretary of Defense

National Defense Research Institute

RAND

Approved for public release; distribution unlimited

The research described in this report was sponsored by the Office of the Secretary of Defense (OSD), under RAND's National Defense Research Institute, a federally funded research and development center supported by the OSD, the Joint Staff, the unified commands, and the defense agencies, Contract DASW01-95-C-0059.

Library of Congress Cataloging-in-Publication Data

Pirnie, Bruce, 1940– .
Assessing requirements for peacekeeping, humanitarian assistance, and disaster relief / Bruce R. Pirnie, Corazon M. Francisco.
p. cm.
"Prepared for the Office of the Secretary of Defense by RAND's National Defense Research Institute."
"MR-951-OSD."
Includes bibliographical references.
ISBN 0-8330-2594-5
1. United States—Armed Forces—Civic action. 2. United States—Armed Forces—Operational readiness. I. Francisco, Corazon. II. United States. Dept. of Defense. Office of the Secretary of Defense. III. National Defense Research Institute (U. S.) . IV. Title.
UH723.P57 1998
355. 3 ' 4—dc21 98-13469
CIP

Published 1998 by RAND
1700 Main Street, P.O. Box 2138, Santa Monica, CA 90407-2138
1333 H St., N.W., Washington, D.C. 20005-4707
RAND URL: http://www.rand.org/
To order RAND documents or to obtain additional information, contact Distribution Services: Telephone: (310) 451-7002; Fax: (310) 451-6915; Internet: order@rand.org

PREFACE

The purpose of this project, sponsored by the Assistant Secretary of Defense (Strategy and Requirements), was to assess requirements for peacekeeping, humanitarian assistance, and disaster relief. The project was carried out in three phases.

During Phase One, RAND was tasked to provide a comprehensive analytic description of events associated with peacekeeping, humanitarian assistance, and disaster relief. To accomplish this task, RAND developed a database, called Force Access, that would be suitable to record and assess these events within the Department of Defense (DoD), Joint Staff (JS), and service staffs, especially the Army Staff. This database includes summary information for relevant operations conducted from 1990 through 1996, lists of units down to battalion/separate-company level for ground forces, and tables that link uniquely identified units to specific operations. Force Access provides a powerful combination of operational history and force structure within an easy-to-use relational database. Fully developed, it will offer an unprecedented look into past operations and a useful tool for exploring the implications for force mix and force structure. An overview of Force Access is given in Chapter Two, and a technical description is contained in Appendix B.

During Phase Two, the sponsor exercised his option to restructure the project. In place of activities originally planned for Phase Two, the sponsor tasked RAND to produce a series of vignettes—generalized descriptions of the types of operations described in Chapter Two—based on operations contained in the Force Access database. These vignettes are presented in Chapter Three. During the same phase, RAND was tasked to analyze the implications of these recurring operations, especially indications of stress on frequently tasked units of various types. This analysis appears in Chapter Four.

During Phase Three, RAND was tasked to recommend changes in force structure and procedures that would improve the conduct of these types of smaller-scale contingencies without detracting from the nation's ability to wage major theater warfare. Such changes include modifications to force mix and force structure across the components. They are enumerated in Chapter Five.

This research was performed within the International Security and Defense Policy Center of RAND's National Defense Research Institute, a federally funded research and development center sponsored by the Office of the Secretary of Defense, the Joint Staff, the unified commands, and the defense agencies. The prospective audience includes decisionmakers and supporting staffs within the Office of the Secretary of Defense and the Joint Staff, but also the services for areas falling within their cognizance.

Comments and inquiries are welcome and should be addressed to the authors (substantive questions: Bruce Pirnie; technical questions: Corazon Francisco).

CONTENTS

FIGURES

TABLES

SUMMARY

PURPOSE OF THE PROJECT

Assessing Requirements for Peacekeeping, Humanitarian Assistance, and Disaster Relief was a three-phase project to assess requirements for such operations and to recommend options to conduct such operations more effectively without detracting from the nation's ability to prevail in major theater warfare. In recent years, U.S. forces have demonstrated that they can conduct these operations very successfully; therefore, large changes are not required.

METHODOLOGY

During the research phase of this project, we constructed a database of unit participation in operations from 1990 through 1996. Although the armed services make little systematic effort to retain such data, we recovered enough to gain a detailed, realistic picture of unit participation, especially for Army forces that were at the heart of the "high-end" operations. *High-end operations* involve ambitious missions (humanitarian intervention and coercive peace operations) and require large forces. Somalia, Haiti, and Bosnia provide examples. We focused on high-end operations because they are the most demanding and the most stressful. Using these data and other sources, such as after-action reports, we composed vignettes that describe operations in a generalized or idealized way. Each vignette includes a mission statement and operational phases. For each phase, we listed the implied tasks and gave a corresponding force structure. For Army and Marine forces, these force structures descend below unit level to detachment and team level. We used these vignettes to discern what types of units would be relatively scarce and, hence, especially stressed under various assumptions about the level of future operations. We enriched this basic research through a secondary-source literature survey and research conducted both within and outside RAND, to develop options that address both force structure and procedures.

RECOMMENDED OPTIONS

The essential U.S. contribution to high-end operations is *power projection,* rapid deployment and sustainment of forces prepared for high-intensity combat. Indeed, U.S. force requirements for these operations and for outright interventions, such as URGENT FURY (Grenada) and JUST CAUSE (Panama), are practically identical. Most of the options proposed in this study would contribute to power projection as broadly defined.

The frequency of these operations is highly uncertain and might decline until they become as rare as during the Cold War. A prudent force planner would prefer to avoid changes that would be unhelpful if operations of this sort were to decline. Most of the options would help accomplish other missions or else would occasion little regret if the operations for which they were designed became rare again.

Refine Command Elements for CJTF

The Joint Staff, unified commands, and services further refine command elements required for a combined joint task force (CJTF). CJTF command elements are organized according to a standard pattern, and their officers are predesignated on a contingency basis.

This option is insensitive to assumptions about the level of future operations, because it entails only modest commitment of resources and would also improve power projection for other purposes, whether conducted unilaterally or in concert with other countries.

CJTF command elements would be activated and exercise frequently enough to ensure that predesignated officers are proficient despite the rotational cycle of normal assignments. During exercises, staffs might assemble in one location, such as a wargaming facility, or they might network from several locations. Moreover, predesignated officers would also become directly acquainted with forces of sister services. Exercises would include play with foreign forces and with those civilian agencies that are frequently involved in humanitarian intervention and peace operations.

Perform Search and Rescue Using a Variety of Forces

Unified commands employ a variety of specialized forces to perform search and rescue in denied areas.

This option would help to even the stress on specialized forces if the level of operations remains at least constant. Search and rescue during air-denial and strike operations puts heavy demands on specialized forces within the active components of the services, especially on certain aircraft (M/HH-60, MC-130, MH-53, HC-130) and their crews. The requirement is inflexible and inescapable because it is determined by level of threat and geographic extent of air operations. To reduce stress on specialized forces, unified commands should spread the burden evenly over all forces appropriate to the mission and the prevailing situation. Army Special Forces and Rangers supported by Army aviation might be appropriate in some situations. In other situations, a Marine Expeditionary Unit—Special Operations Capable might be appropriate.

Expand Use of Civilian Contractors

The services expand use of civilian contractors to support contingency operations ranging from disaster relief to coercive peace operations.

This option would be appropriate when time permitted to employ civilians and they would not be subjected to excessive risk. It is insensitive to the level of operations; it would merely be exercised less frequently if the level declines.

The Army's Logistics Civil Augmentation Program (LOGCAP) and comparable programs in sister services do not obviate requirements for support units that perform functions similar to those performed through LOGCAP. The services still require military support units that will continue to perform their duties under conditions that would be intolerable for civilian organizations. But peace operations usually imply less-demanding conditions that allow use of civilian contractors, even during initial phases. Use of contractors lessens call-up of inactive components and can save money by employing local labor at a rate much lower than U.S. active-duty pay.

Define Structure of Army Contingency Brigades

The Army defines the structure of those Army contingency brigades that would be activated when the need arose.

This option would do no harm if the level of operations declines; it would be highly beneficial if the level remains at least constant. At little cost, it makes Army forces a more effective and better-understood instrument of national power in a broad range of contingencies, including unilateral interventions. The Army could gain the following advantages:

- Army forces would deploy quickly on short notice.
- Army forces would operate more efficiently, especially during the critical first phase of an operation.
- National Command Authority and JTF commanders would better understand and appreciate Army capabilities.

Under the current organization, the brigade is an operational level of command that requires support (administrative, logistics, medical, etc.) from division and corps. If a brigade has to operate independently, slices of higher-echelon assets must be provided. Providing these on an *ad hoc* basis causes turmoil and initial uncertainty for the officers controlling the operation. (An exception is the ready brigade of 82nd Airborne Division, which routinely prepares for independent operations.) Of course, exigencies of different theaters and missions may require variations in the slices, but recent practice indicates that the broad requirements are well understood and fairly stable. Organizing contingency brigades with integral support would increase efficiency, especially during deployment and the initial phase of operations.

Use National Guard and Reserve for Noncoercive Peacekeeping

As a matter of policy, the Army uses National Guard and Reserve forces for noncoercive peacekeeping.

This option is almost insensitive to assumptions about the level of future operations. At a diminished level, it tends to become irrelevant; at a constant level or higher levels, it becomes worthwhile.

Traditional peacekeeping is well within the capability of National Guard and Reserve units and individual members. In this type of operation, a military force monitors compliance with an agreement, reports violations, and attempts to resolve violations, usually through negotiations with representatives of the parties. This military force is not expected to enter combat, except in self-defense if attacked.

Employment of National Guard and Reserve forces would offer two main advantages: less diversion of active units (currently two infantry battalions), making them more available for more-important contingencies and major theater warfare, and no loss of training opportunities by active units in order to prepare for peacekeeping, conduct it, and reconstitute afterwards. Conversely, National Guard and Reserve forces would improve their abilities to mobilize, deploy, and conduct operations requiring basic soldiering skills. Such employment would also have disadvantages: increased cost, largely due to the difference between normal pay and active-duty pay; and more active duty away from home, a disadvantage that could be mitigated by preferring volunteers.

Develop Modularity Below the Unit Level

The Army and Air Force continue to develop modularity below unit level.

This option would entail little additional expense and, therefore, few regrets if the level of operations declines. In division- and corps-sized operations, deployment is normally by unit for the obvious reason that units are organized to conduct operations on this scale. But in the smaller deployments thus far characteristic of the period since the Cold War, deployment has been brigade-sized and smaller, causing much fragmentation of units. Flexibility is not an issue. The Army could hardly be more flexible in its willingness to task-organize forces. Indeed, modularity tends to limit flexibility by offering preconceived entities to the planner. But it may be advisable to sacrifice some flexibility in order to gain smoother, more predictable execution.

Modularity has least relevance to combat units, because they are already fungible to the lowest levels compatible with effective combat. It has the greatest relevance to support units that have been increasingly required to perform discrete portions of the overall missions of such units. Implicit in modularity is an understanding of the transportation required to deploy, of sustainment in the field, and of doctrinal statements of the capabilities and limitations of these modules.

Increase Readiness of Selected Army Support Units

The Army increases the readiness of frequently deployed and relatively scarce support units.

This option is moderately sensitive to assumptions about the level of future operations. If the level remains at least constant, this option would be advisable; if the level declines, then the Army would incur additional expense without commensurate gain.

Army units are accorded authorized levels of organization (ALO)[1] in accordance with anticipated requirements. Support units outside the maneuver divisions are generally accorded lower ALO, as are many units in the Reserve and National Guard. If time permits, units at ALO 1 are usually refreshed prior to deployment. During the Cold War, many types of support units were required at such infrequent intervals that it was sound policy to conserve resources by keeping them less ready. But since the end of the Cold War, frequent smaller-scale contingencies have upset this calculation. At an increased operational tempo, there are fewer opportunities to conserve resources

[1]ALO are authorized levels of personnel and equipment, expressed as percentages of Tables of Organization and Equipment (TOE) or Modified Tables of Organization and Equipment (MTOE) levels. ALO 1 is 98 to 100 percent; ALO 2 is 88 to 97 percent; ALO 3 is 78 to 87 percent of TOE/MTOE levels. For example, early-deploying units (Force Package 1) are normally allocated ALO 1.

because units have to be made ready anyway before they deploy. Moreover, when less-ready units are called up, they usually have to draw personnel and equipment from sister units ("cross-leveling"), causing turbulence and further reducing readiness in the losing units—perhaps to the point that they can no longer train effectively.

Selection of units would depend on multiple factors, including complementary types (e.g., topographical companies at various echelons), current readiness levels, geographic locations, wartime missions, scheduled changes in Table of Organization and Equipment (TOE) and Modified Table of Organization and Equipment (MOTE), and relative scarcity across the components.[2] For example, the Army could improve military police support by increasing readiness among the large pool of military police units in the active component. By contrast, civil affairs resides almost entirely in the inactive components; there is just one civil affairs battalion in the active Army. A decision to increase the readiness of a Reserve or National Guard unit would not necessarily imply that it would deploy abroad: It might be slated to replace an active unit that deployed, a process called "backfilling."

Add Support Units to the Active Army

The Army adds some frequently deployed low-density support units to the active component.

This option is highly sensitive to assumptions about the level of operations. If that level remains at least constant, this option would be advisable; if the level declines, then the Army would regret sacrificing other priorities to add support units that found little employment. The Army would especially regret sacrificing combat power if combat units had to be traded for noncombat units within end-strength.[3]

Following the Vietnam War, the Army deliberately moved support units to the Reserve, a move that allowed the Army to maintain greater combat power than would otherwise have been possible

---

[2]TOE/MTOE authorize personnel and equipment for types of units, e.g., a light-infantry company.

[3]*End-strength* is the limit set by legislation on the personnel strength of military forces.

within a constrained budget. The active component kept a high proportion of combat units and enough support units to initiate large-scale operations. The reserve component acquired enough support units to sustain large-scale operations. This division of labor has functioned well even at the high level of operations experienced in recent years. Individuals and units from the inactive components have performed competently, and no retention problems have yet emerged. But it is not clear whether National Guard and Reserve could sustain the current level of operations indefinitely.

Develop Air Expeditionary Forces for Close Air Support

The Air Force develops air expeditionary forces optimized to provide the close air support often required during humanitarian intervention and coercive peace operations.

Such expeditionary forces would be equally useful in other operations of comparable size, including unilateral interventions. Since the Air Force has already developed the Air Expeditionary Force (AEF) concept and has repeatedly deployed such forces, this option would probably entail little additional expense.

RESTORE HOPE/CONTINUE HOPE (Somalia) and JOINT ENDEAVOR/JOINT GUARD (the former Yugoslavia) have required close air support of land forces. To provide such support, an AEF would need forward air controllers, both airborne and on the ground, plus a systems-and-munitions mix optimized for the mission. In addition, it would need a command element, probably including an airborne command post, that would be responsive to requests from land forces.

Make Increased Use of Unmanned Aerial Vehicles

The Air Force promotes development and use of unmanned aerial vehicles (UAVs) to diminish the demand for manned platforms in reconnaissance, electronic warfare, and other missions.

This option is insensitive to assumptions about the level of operations considered in this report. Unmanned platforms are highly desirable in a wide range of situations and should be developed even if the level of these operations declines.

SUMMARY OF OPTIONS

Most options are fairly insensitive to the level of operations. If the types of operations described in this report were to decline in frequency and size, most options would remain desirable or at least unobjectionable. (See Figure S.1.) The two exceptions both concern Army units:

- Increasing the readiness of support units, whether in the active or inactive components, would yield benefits only if operations remained at least at the current level.
- Adding support units to the active component might be counterproductive if the level declined. Anticipating problems that have

RANDMR951-S.1

Option	Agencies	Sensitivity to Level of These Operations	Help Improve Power Projection?
Refine command elements for combined joint task forces.	Joint Staff, unified commands, services	Low	Yes
Use a variety of specialized forces to perform search and rescue.	Unified commands, services	Low	—
Expand selective use of civilian contractors.	Unified commands, services	Low	Yes
Organize Army contingency brigades during peacetime.	Army	Low	Yes
Use National Guard and Reserve for noncoercive peacekeeping.	Army	Low	—
Develop modularity below the unit level.	Army, Air Force	Low	Yes
Increase readiness of selected Army support units.	Army	Medium	Yes
Add support units to the active Army.	Army	High	—
Develop Air Expeditionary Forces optimized for close air support.	Air Force, Army	Low	Yes
Make increased use of unmanned surveillance platforms.	Services	Low	Yes

Figure S.1—Recommendations

not yet fully emerged, the Army would have allocated resources to support units that might better have been allocated to modernization, among other pressing needs.

Most of the options would not only improve conduct of humanitarian interventions and coercive peace operations, they would also improve power projection for other purposes, including unilateral interventions such as JUST CAUSE (Panama) and U.S.–led multilateral operations such as URGENT FURY (Grenada). Three options stand out as especially helpful in this broader context:

- Further development of command elements for combined joint task forces
- Organization of Army contingency brigades
- Development of air expeditionary forces optimized for close air support.

Contingency brigades and air expeditionary forces are a natural fit with strong synergistic effects: a powerful, versatile force appropriate for a wide range of contingencies.

ACKNOWLEDGMENTS

We gratefully acknowledge the assistance of many persons, civilian and military, outside RAND: MAJ Steve Aviles, U.S. Army Concepts Analysis Agency (CAA), provided deployment data for Joint Task Force (JTF) Andrew, which had been produced originally by the Army Operations Center (AOC). COL Michael D. Angelo, U.S. Army Reserve, and Col. Fred L. Baker, U.S. Air Force Reserve, assigned to the Office of the Assistant Secretary of Defense (Reserve Affairs), advised on employment of National Guard units in peacekeeping. COL Benton H. Borum, Chief, Force Readiness Division, Readiness and Mobilization Directorate, Army Staff, provided data on U.S. Army deployments and advice on certain recommendations, especially the proposed Army Contingency Brigade. CDR Paul Braszner, Forces Division, J-8, gave advice on typical employment of Navy forces. LtCol G. M. Doermann and Elma Barber, Operations Division (POC-30), Headquarters, U.S. Marine Corps, provided Status of Resources and Training System (SORTS) data for U.S. Marine Corps units. Edward Drea, Chief, Research and Analysis Division, U.S. Army Center of Military History, provided research support. COL Larry M. Forster, Director, U.S. Army Peacekeeping Institute, U.S. Army War College, and Walter S. Clark commented on the emerging recommendations.

Duane Gory served as a point of contact at CAA and organized a useful meeting to discuss available materials. CAPT Don Hutchins and CAPT Jack Picco, Seabee Program, Facilities and Engineering Division (N-44), Deputy Chief of Naval Operations (Logistics), provided data on Seabee operations. Maj. Steve Kempf, Forces Division, J-8, and CAPT Thomas J. Gregory, Head, Force Assignment Branch,

J-8, provided deployment data collected pursuant to the Quadrennial Defense Review (QDR). Renee Lajoie, Defense Forecast Incorporated (DFI) International graciously shared an extensive database of U.S. Air Force Operations. LtCol Colin D. Lampard, a fellow with RAND, advised on Marine Corps operations and force structure. John Runkle, Head of the Force Builder Team, U.S. Army Force Management Support Agency, a field operating agency reporting to the Deputy Chief of Staff, Plans and Operations (ODCSOPS), provided data for U.S. Army units pursuant to the Quadrennial Defense Review. With all transactions (except deactivation) deleted, these data form the basis for Army force structure depicted in Force Access.

Frank N. (Micky) Schubert, Joint Staff Historical Office, Office of the Chairman of the Joint Chiefs of Staff, provided access to unified command histories. Adam B. Siegel, Center for Naval Analyses (CNA), provided insights from his experience and hosted a useful conference concerning military operations other than war (MOOTW) on September 25, 1996. Steven B. Siegel, Chief, Resource Analysis Division, U.S. Army Concepts Analysis Agency, provided an overview of relevant work accomplished by CAA. Wayne Thompson, Air Force History Support Office, provided data on DELIBERATE FORCE (the former Yugoslavia) drawn from his research. William E. Stacy, Command Historian, U.S. Forces Command (USFORSCOM), and Warner Stark, assigned in the Office of the Command Historian, provided deployment data for recent operations.

We extend thanks to numerous RAND colleagues: John Bordeaux and Jennifer Kawata offered invaluable assistance in processing raw data; Glenn Gotz and Kathi Webb shared research conducted in the course of a wider project examining U.S. overseas presence; Paul Killingsworth provided Air Force data and advised on employment of air forces; Maren Leed provided a wealth of material and contacts during her stay at the Pentagon. Leslie Lewis and Roger Brown provided invaluable assistance to this project in the context of QDR support. Susanna Purnell conducted basic research for the Force Access database. Jennifer Taw, Ronald Sortor, and David Persselin shared their analysis of readiness issues and deployment data. Jennifer Ingersoll-Casey provided research support. David Thaler shared his calculations of Air Force units required to sustain the current tempo of operations and still meet the goal of no more than 120 days' temporary duty per year. Alan Vick shared his analysis of Air Force requirements for contingency operations. Luetta Pope did an excellent job of preparing the report for publication, despite technical difficulties. Marian Branch edited the final draft with exceptional skill and attention to detail. Betty Amo handled the galley proofs.

We extend thanks to our project monitor Matt Vaccaro and also to Frank Jones, Office of the Assistant Secretary of Defense (Strategy and Requirements), for supporting our research and guiding development of Force Access. We are indebted to Charles L. Barry for conducting an exceptionally thorough and expert review of the draft. Had time permitted, we would have added an analysis of space assets in support of peace operations, as Chuck recommended in his review: "Today we rely on space for intelligence, telephone and radio communications, video conferences, weather data and the now-essential internet. The need for space systems and ground links is no less critical than ships and ground forces. A comprehensive study should include this facet of contingency operations."

ABBREVIATIONS AND ACRONYMS

AC	coercive humanitarian assistance
AE	ammunition ship
AEF	Air Expeditionary Force
AFB	Air Force Base
AFR	Air Force Reserve
ALO	authorized level of organization
AMC	Air Mobility Command (U.S. Air Force)
AMC	Army Materiel Command
AN	noncoercive humanitarian assistance
ANG	Air National Guard
ANGLICO	air and naval gunfire liaison company
AO	area of operations
AOC	Army Operations Center
ARG	amphibious ready group
Bn	battalion
CAA	U.S. Army Concepts Analysis Agency
CARE	Co-operative for Assistance and Relief Everywhere
CARRS	Combat Arms Regimental System
CAS	close air support
CBT	combat
CIS	Commonwealth of Independent States
CJTF	combined joint task force (NATO usage)
CNA	Center for Naval Analyses
CONUS	continental United States
COSCOM	Corps Support Command

CPX	command post exercise
CS	combat support
CSS	combat service support
DCSOPS	Deputy Chief of Staff, Plans and Operations
Det	detachment
DFI	Defense Forecast Incorporated
DoD	Department of Defense
DR	disaster relief
DS	direct support
EAC	echelons above corps
EOD	explosive ordnance disposal
EPW	enemy prisoners of war
FA	Field Artillery
FAC	forward air controller
FTX	field training exercise
FWE	fighter wing equivalent
GS	general support
GTMO	Guantánamo (Cuba)
HARM	High-Speed Antiradiation Missile
HHB	headquarters and headquarters battery
HHC	headquarters and headquarters company
HHD	headquarters and headquarters detachment
HQ	headquarters
Hvy Div	Heavy Division
IN	Infantry
Inf Div	Infantry Division
JS	Joint Staff
JTF	joint task force
LAD	latest arrival date
LANTIRN	Low-Altitude Navigation and Targeting Infrared for Night
LHA	Landing Helicopter Assault [Ship]
LHD	Multipurpose amphibious assault ship
LMI	Logistics Management Institute
LOGCAP	Logistics Civil Augmentation Program
LPD	Landing Platform—Dock
LSD	Landing Ship—Dock

Lt Div	Light Division
MAGTF	Marine air-ground task force
MASH	Mobile Army Surgical Hospital
MB	megabyte
MEB	Marine Expeditionary Brigade
MEF	Marine Expeditionary Force
MEU	Marine Expeditionary Unit
MEU (SOC)	Marine Expeditionary Unit—Special Operations Capable
MI	Military Intelligence
MOG	maximum [number of aircraft] on ground
MOMEP	Military Observer Mission in Ecuador and Peru
MOOTW	military operations other than war
MP	Military Police
MSCA	military support to civil authorities
MSE	multiple subscriber equipment
MTOE	Modified Table of Organization and Equipment
NATO	North Atlantic Treaty Organization
NCA	National Command Authority
NGO	nongovernmental organization
NMCB	Naval Mobile Construction Battalion
OCONUS	outside the continental United States
OPS	operations
PAA	primary aircraft authorized
PAF	Pacific Air Forces (U.S. Air Force)
PC	coercive peace operations
PLS	Palletized Load System
PN	noncoercive peace operations
PRIME BEEF	Prime Base Engineer Emergency Force
Pysch Ops	Psychological Operations
QBE	Query-by-Example
QDR	Quadrennial Defense Review
QRF	quick-reaction force
RAM	random access memory
RDBMS	relational database management system
RED HORSE	Rapid, Engineer-Deployable, Heavy Operational Repair Squadron, Engineer

SAIC	Science Applications International Corporation
SAG	surface action group
SAMAS	Structure and Manpower Allocation System
SAR	search and rescue
SEAL	Sea, Air, Land
SO	special operations
SOC	Special Operations Capable
SORTS	Status of Resources and Training System
SP	self-propelled
Spec Ops	Special Operations
SRC	Standard Requirements Code
Strat Dissem	Strategic Dissemination
TACSAT	Tactical Communications Satellite
TALCE	Tanker Airlift Control Element
TDA	Table of Distribution and Allowances
TDY	temporary duty
TPFDD	Time Phased Force Deployment Data
TOE	Table of Organization and Equipment
Tropo	Tropospheric
UAV	unmanned aerial vehicle
UIC	Unit Identification Code
UN	United Nations
UNAVEM II	Second United Nations Angola Verification Mission
UNIKOM	United Nations Iraq-Kuwait Observation Mission
UNMIH	United Nations Mission in Haiti
UNOSOM I	First United Nations Operation in Somalia
UNOSOM II	Second United Nations Operation in Somalia
UNPROFOR	United Nations Protection Force
UNSC	United Nations Security Council
UNTAC	United Nations Transitional Authority in Cambodia
USA	U.S. Army
USACOM	United States Atlantic Command
USAF	U.S. Air Force
USAFR	U.S. Air Force Reserve
USEUCOM	United States European Command

USFORSCOM	United States Forces Command
USMC	U.S. Marine Corps
USN	U.S. Navy
USPACOM	United States Pacific Command
USSOCOM	United States Special Operations Command
USSOUTHCOM	United States Southern Command
USTRANSCOM	United States Transportation Command
UTC	unit-type code

Chapter One

INTRODUCTION

The purpose of this study was to assess requirements for peace operations, humanitarian assistance, and disaster relief, then to develop options to conduct such contingencies more effectively without detracting from the nation's capability to conduct major theater warfare. This report focuses on one aspect of requirements: those military units required to accomplish the particular missions.

TREND SINCE 1989

Since the collapse of Soviet power in 1989, U.S. forces have been engaged to an unprecedented extent in smaller-scale contingencies. (See Appendix A for a list of the contingencies considered in this study.) Most of these can be subsumed under the broad categories of "humanitarian assistance," "humanitarian intervention," and "peace operations." Most instances of humanitarian assistance are occasioned by recurring natural disasters, which are predictable in the aggregate. With some exceptions, this assistance demands only limited military forces for days or weeks. Humanitarian intervention is occasioned by conflict, a less predictable phenomenon that can demand extensive military forces for months and even years. Peace operations cover a wide spectrum, from traditional peacekeeping (monitoring and reporting activity) to peace enforcement, a far more demanding mission requiring highly capable combat forces.

CONCERNS CAUSED BY THE TREND

The recent trend toward more-frequent military engagement in small operations has occasioned concerns, including the following:

- Does the U.S. military have sufficient units of the appropriate types to conduct these operations without undue stress?
- Do these operations reduce capability to an unacceptable degree by degrading combat skills, wearing out equipment, and making units less ready to deploy?
- Is the increased frequency of deployment having unacceptable effects on morale and retention of military personnel?

This study focuses on the first of these concerns. This concern arises because the U.S. military was developed to wage major theater warfare against the Soviet Union and its Warsaw Pact allies, not to conduct the types of operations considered in this report. With regard to combat, the required capabilities overlap: U.S. forces that could have defeated Soviet forces can easily overwhelm Somali clans or Bosnian Serb militia. But smaller-scale contingencies place especially heavy demands on combat support and combat service support, units that may not be sufficiently ready or easily accessible. They also place especially heavy demands on certain specialized forces such as units that conduct search and rescue. Moreover, protracted operations require rotation, usually in a 6-month cycle, which requires multiple units of similar type. Changes in force structure or procedures may be advisable to ensure that the appropriate units are available in sufficient numbers.

METHODOLOGY

To assess requirements for number of military units, we used an historical method. We researched contingency operations conducted from 1990 through 1996 to discern the missions and to develop lists of military units that participated. We entered these lists into a relational database that allowed us to make queries based on types of operations or types of units. With a few notable exceptions, especially the later period of operations in Somalia, we assumed that those units that participated were also required to perform the mission. Of course, this assumption is debatable. In some cases, the joint task force may have deployed more or fewer units than were actually required or deployed some units of the wrong types. It is particularly difficult to assess how much force may have been required to appear overwhelming in the eyes of potential adversaries. Nevertheless, we believe that this assumption is sufficiently valid to generate a useful study of requirements.

We abstracted from the historical operations to develop vignettes of typical operations. These vignettes assist analysis by reducing the messy, quirky events of actual operations to a more intelligible pattern. We used these vignettes to examine requirements, both peak strength and rotational demands, under broad projections of the level of future operations. Finally, we combined this analysis with results of other studies, both within and outside RAND, to develop options that would improve capability to conduct these operations successfully. We especially emphasized options that would be advisable even if the level of operations declined, possibly to that of the Cold War.

ORGANIZATION OF THIS REPORT

In Chapter Two, we review the history of operations during the period of interest. We analyze the types of operations and, for each type, we assess frequency, duration, and level of effort, expressed in military units. At the conclusion of the chapter, we summarize these operations and broadly evaluate outcomes to determine whether the missions were accomplished successfully.

In Chapter Three, we develop vignettes that are intended to serve as archetypes for purposes of analysis. At the conclusion of the chapter, we summarize the factors that determine force requirements.

In Chapter Four, we analyze implications at various levels of future operations and for the forces of all armed services. We emphasize particularly those Army units that are central to protracted land operations and those Air Force units that are required to secure no-fly zones and conduct strikes.

In Chapter Five, we recommend options that would improve capability. For the most part, these are changes or adjustments at the margin, because U.S. forces have clearly demonstrated that they have sufficient capability to perform these operations successfully. We emphasize the benefit to be gained by defining the structure of Army contingency brigades, which would be activated when the need arose, and combining these brigades with suitably tailored air expeditionary forces.

Chapter Two

OVERVIEW OF OPERATIONS, 1990–1996

"Force Access," the database we developed during Phase One of the project, supports an overall assessment of requirements for peacekeeping, humanitarian assistance, and disaster relief. This chapter gives an overview of operations from 1990 through 1996 that are contained in that database. (See Appendix B for a technical description of Force Access.)

The database and associated analysis consider forces that are deployed or directly employed, including, for example, forces that accomplish disaster relief near their home stations. They do not consider forces that remain at their home stations and are not directly employed but direct much of their effort to supporting other forces that are involved directly. Contingency operations demand many supporting efforts that are not easily captured. Some efforts are sporadic—for example, support at ports of embarkation and debarkation during an initial deployment. Other efforts are continuous—for example, the flow of sustaining supplies from home bases to the area of operations. In many instances, it is difficult to assess how much additional effort is occasioned by contingency operations as distinguished from normal day-to-day operations that must be supported anyway.[1]

---

[1]In his review of the draft report, Charles Barry offered these distinctions: "First, the impact on the workload of units that continue to perform their routine duties at the same base, but who have altered how they do it and whom they do it for. Second, what forces are so affected that the operation has become a major consumer of their time, personnel and resources. . . . These may be units that have themselves deployed for the operation or have simply been tasked for high priority or dedicated support."

CONUS HUMANITARIAN ASSISTANCE

CONUS (continental United States) humanitarian assistance operations involve support by military forces of civil authorities within the contiguous 48 states following those natural disasters that briefly overwhelm the ability of civil authorities to respond. Such disasters include storms, floods, forest fires, earthquakes, and catastrophic accidents. In addition, military forces may initially assist in restoring civil order.

Frequency

U.S. forces may be called upon at any time to provide limited humanitarian assistance on a local level, almost always in the form of disaster relief. Some of this assistance occurs so frequently and at such a low level as to fall beneath the threshold of this report. For example, the National Guard frequently assists at the local level and the Coast Guard constantly offers assistance to mariners in distress.

U.S. forces respond to certain types of natural disasters that recur quite predictably. Each fall, hurricanes threaten the southeastern United States. About every other year, one or more of these storms cause enough damage to require military assistance. With about the same frequency, spring thaws and heavy rains can cause flooding, especially in the Missouri, Ohio, Mississippi, and Sacramento River valleys. Dry summers and falls lead to dangerous forest fires in the northwestern United States and California. Earthquakes are extremely common along the California fault lines but require military assistance infrequently. Overall, U.S. forces conduct operations providing humanitarian assistance within CONUS 1–3 times annually.

Duration

CONUS humanitarian assistance usually lasts 2–14 days. An exceptionally ruinous storm may demand longer efforts. For example, the joint task force formed in response to Hurricane Andrew remained in existence for 52 days (August 25–October 15, 1992). As another exception, a wide debris field and difficult diving conditions caused search efforts following a 1996 Transworld Airways crash off Long Island to last from July into November.

Level of Effort

CONUS humanitarian assistance usually requires small joint task forces (JTFs) that may include fixed-wing airlift, helicopters, amphibious ships, Coast Guard cutters, naval construction battalions, and various Army and Marine units. These JTFs are typically tasked to assess damage. They provide search and rescue, emergency communications, electrical power, medical care, food, potable water, and shelter. They may also remove debris and sometimes lift grounded ships (for example, following Typhoon Omar, August–September 1992).

Since 1990, the highest level of effort has been for JTF Andrew, which was built around elements of the 10th Mountain Division and XVIII Airborne Corps and was supported by extensive airlift and some Navy and Marine support. At peak, almost 22,000 Army troops drawn from at least 84 units participated in this operation.

Higher Level of Effort

Previous efforts do not set the upper limit for CONUS humanitarian assistance. There might be a major catastrophe within the United States, necessitating a far larger relief operation than has been experienced previously. Hurricane Andrew would have been far more destructive had it passed through the greater Miami area, and would have required a commensurately larger operation in response. As other examples, California could experience massive movement along the notorious San Andreas Fault ("the big one") or the south-central United States could be devastated by movement along the Madrid Fault. In these cases, the ensuing disaster relief operation could be an order of magnitude greater than that required for Andrew.

OCONUS HUMANITARIAN ASSISTANCE

In OCONUS (outside the continental United States) operations, military forces render humanitarian assistance to civil authorities in areas outside the contiguous 48 states following those natural disasters that briefly overwhelm the ability of civil authorities to respond. In Alaska, Hawaii, and U.S. territories, U.S. forces may initially assist in restoring civil order. In foreign countries, the local authorities are normally responsible for restoring civil order.

U.S. military forces provide humanitarian assistance outside the continental U.S. when civil authorities in Alaska, Hawaii, U.S. territories, and sometimes in foreign countries are temporarily overwhelmed by disasters such as typhoons, hurricanes, droughts, famines, epidemics, oil fires, earthquakes, catastrophic accidents, and dislocation caused by political turmoil, as in the former Soviet Union. Very often, such assistance is limited to airlift of equipment and supplies.

Frequency

Not counting brief uses of airlift, U.S. forces usually provide OCONUS humanitarian assistance 1–3 times per year. The most frequent type of operation is relief following typhoons, which regularly inflict damage along the Pacific Rim, throughout the South Pacific, and in the Caribbean.

Duration

Most responses have lasted 2–4 days, but some operations have extended beyond that. For example, FIERY VIGIL (evacuation of personnel and dependents from the Philippines following the eruption of Mount Pinatubo) took 22 days (June 8–30, 1991). SEA ANGEL (disaster relief to Bangladesh following a typhoon) took 34 days (May 11–June 13, 1991). Both PROVIDE HOPE (the former Soviet Union, April 3–July 24, 1992) and PROVIDE RELIEF (assistance to refugees in Somalia, August 15–December 2, 1992) lasted over three months, but they were almost exclusively airlifts.

Level of Effort

Because of limited tasks and restrictive rules of engagement, humanitarian assistance usually requires only small ground forces. For example, SUPPORT HOPE (Rwanda) included a Marine infantry company (A Company, Battalion Landing Team 1/4) and an Army airborne infantry company (C Company, 3–325 Infantry). Noncombat Army forces in this operation included elements of two transportation battalions, elements of a support company, a signal detachment, an engineer platoon, an ordnance team, a water-purification team, and a preventive-medicine detachment.

Humanitarian assistance can demand large numbers of aircraft missions. PROVIDE HOPE (the former Soviet Union) totaled 700 missions. PROVIDE RELIEF (Somalia) totaled 3,094 missions, an average of 28 missions per day.

SEA ANGEL was an exceptional operation. It was possible because the 5th Expeditionary Brigade (MEB) was fortuitously under way from the Persian Gulf following the war there. This operation was accomplished primarily by elements of the 5th MEB and Naval Task Group 76.6, plus elements of the Air Force's 374th Tactical Airlift Wing, the Army's 4/25 Aviation Battalion, and small numbers of special operations forces.

MILITARY SUPPORT TO CIVIL AUTHORITIES

In the context of this report, military support to civil authorities (MSCA) includes all support other than humanitarian assistance and counterdrug operations rendered to civil authorities in the United States and its territories.

Frequency

Aside from humanitarian assistance and counterdrug operations, the only MSCA since 1990 has been GARDEN PLOT (Joint Task Force Los Angeles), conducted in May 1992 to restore civil order during the riots following the first trial of police officers accused of beating Rodney King.

Duration

GARDEN PLOT lasted 13 days (May 1–12, 1992).

Level of Effort

Three Army heavy brigades from the California National Guard were alerted for GARDEN PLOT, but the riots subsided before substantial assistance was required.

OPERATIONAL AIRLIFT

This category includes military airlift of personnel and equipment in support of military intervention or peace operations. The United States frequently airlifts forces controlled by the U.N. and nationally controlled forces of its allies in such operations.

Frequency

Since 1990, the United States has airlifted allied forces on one occasion: 600 French troops into the Central African Republic on February 26–27, 1991. On six other occasions (IMPRESSIVE LIFT I and II [Pakistani forces to Somalia], QUICK LIFT [allied rapid-reaction force to Croatia], other UNPROFOR [U.N. Protection Force] lift, lift in support of U.N. peace operations in Rwanda, the Second United Nations Operation in Somalia [UNOSOM II] lift, the United Nations Transitional Authority in Cambodia [UNTAC] lift), it airlifted personnel and equipment committed to U.N. operations. On the basis of these precedents, we figure that the U.S. Air Mobility Command (AMC [Air Force]) is likely to provide operational airlift 1–2 times per year.

Duration

Most examples of operational airlift have been accomplished in one to three weeks. For example, QUICK LIFT, transport of the Anglo-French Rapid Reaction Force to Croatia, took five days (June 30–July 4, 1995).

Level of Effort

The level of effort has varied from several sorties to larger efforts, such as QUICK LIFT, which involved 80 missions lifting 4,700 passengers and 1,504 tons of equipment and supplies.

MIGRANTS

The United States has employed military force to intercept and detain foreign nationals attempting to enter this country illegally.

Frequency

Since 1990, the United States has conducted six operations to prevent illegal migration to the United States from China, Cuba, and Haiti (SAFE HARBOR, DISTANT HAVEN, ABLE VIGIL, PROVIDE REFUGE, PROMPT RETURN, and SAFE HAVEN/SAFE PASSAGE). In two operations (PROVIDE REFUGE [1993], PROMPT RETURN [1995]), Chinese nationals were attempting to flee their country. The other four operations involved Cubans and Haitians. Projecting this rate into the future, the U.S. Navy and Coast Guard could expect to conduct one operation of this type each year.

Duration

Considered individually, these operations have lasted months to years. The longest has been SAFE HARBOR (also known as JTF GTMO), which lasted about 20 months (November 1991 to June 1993). But if SAFE HARBOR, ABLE VIGIL, DISTANT HAVEN, and SAFE HAVEN/SAFE PASSAGE are considered phases of one protracted operation, the duration exceeds three years (November 1991 to February 1995).

Level of Effort

These operations typically have involved a small surface action group, composed of several destroyers and frigates, an amphibious ready group (ARG), and several Coast Guard cutters. They also have required small numbers of Army and Marine Corps units to screen, control, house, feed, and otherwise care for tens of thousands of Cubans and Haitians at camps in Guantánamo, Panama, and Surinam.

SANCTIONS

The United States has enforced sanctions declared by the United Nations Security Council (UNSC), primarily by interdicting maritime traffic.

Frequency

Since 1990, the United States has conducted four operations to enforce UNSC sanctions. Beginning in August 1990, Maritime Interception Operations were conducted by a multinational force to enforce sanctions against Iraq. SEA SIGNAL/SUPPORT DEMOCRACY was conducted in the Caribbean to enforce economic sanctions against the Cedras regime in Haiti. MARITIME MONITOR and SHARP GUARD/MARITIME GUARD were conducted in the Adriatic to enforce a prohibition on arms shipments to states of the former Yugoslavia. Given this record, U.S. forces might have to conduct up to two such operations annually.

Duration

Sanctions are likely to extend for months and years. SEA SIGNAL/SUPPORT DEMOCRACY lasted 11 months (October 18, 1993–September 19, 1994) and had little apparent effect on the Cedras regime. MARITIME MONITOR, conducted without coercion, and its successor, SHARP GUARD/MARITIME GUARD, conducted forcibly, lasted over four years (July 1992–October 1996.)

Level of Effort

Enforcement of sanctions usually requires maritime surveillance aircraft and a surface action group. It may also require boarding parties (Sea, Air, Land troops [SEALs], specially trained Marines, and Coast Guard personnel) to take control over ("take down") ships that respond unsatisfactorily to challenge.

TRADITIONAL PEACEKEEPING

Traditional peacekeeping implies observation and monitoring by military forces, normally under the authority of resolutions of the United Nations Security Council. The United States has participated in three traditional peacekeeping operations since 1990 (Multinational Force, ABLE SENTRY, SAFE BORDER). Prior to the collapse of European Communism in 1989, great powers usually did not participate in peacekeeping because Cold War rivalries would have mitigated against their impartiality. As an exception, the United States participated in the Multinational Force. (The United States would have preferred to send a U.N. force, but the Soviet Union, angered by the Camp David Accords and its consequent loss of influence, blocked action in the Security Council.)

Frequency

After 1989, participation of the great powers became less objectionable and the United States has participated in two other peacekeeping operations (ABLE SENTRY, SAFE BORDER).

Duration

Traditional peacekeeping lasts months and years. The more important operations tend to become open-ended. SAFE BORDER (support to the Military Observer Mission in Ecuador and Peru) lasted 16 months (March 1995–June 1996). The Multinational Force (monitoring certain areas of the Sinai Peninsula), established in April 1982, sees no end in sight, nor does ABLE SENTRY (monitoring the northern border of Macedonia), established in July 1993.

Level of Effort

Usually, traditional peacekeeping is built around light and mechanized infantry battalions. The United States currently deploys one light infantry battalion each in the Sinai and in Macedonia; these battalions are supported by elements of rotary-wing aircraft.

NO-FLY ZONE

The United States has employed military force to establish no-fly zones under the authority of the United Nations Security Council and declarations consonant with these UNSC resolutions.

Frequency

Since 1990, there have been essentially three no-fly operations: PROVIDE COMFORT II/NORTHERN WATCH (Northern Iraq), SOUTHERN WATCH (Southern Iraq), and DENY FLIGHT/DECISIVE EDGE (former Yugoslavia, especially Bosnia-Herzegovina).

Duration

No-fly operations have taken years and have tended to become open-ended. DENY FLIGHT (Bosnia-Herzegovina) lasted over two years (April 12, 1992–December 19, 1995). During DENY FLIGHT, U.S. and other air forces were also tasked to provide close air support and conduct punitive strikes as required. A large-scale air strike was conducted to protect "safe areas" declared by the Security Council (DELIBERATE FORCE/DEADEYE) and lasted 23 days (August 29–September 20, 1995). DECISIVE EDGE, the follow-on operation in support of the NATO-controlled Implementation Force, is currently in progress.

No-fly operations over Iraq have already lasted over five years, and there is no end in sight. PROVIDE COMFORT II/NORTHERN WATCH, begun in July 1991, was initially intended to protect the Kurdish population in northern Iraq. SOUTHERN WATCH, begun in August 1992, was initially intended to protect the Shi'ite population of southern Iraq. No-fly operations have not been very effective in protecting either group, but they do help keep pressure on the Iraqi regime to cooperate with the U.N. commission investigating the regime's projects to develop weapons of mass destruction.

Level of Effort

No-fly operations usually have required one U.S. composite wing and several squadrons of allied aircraft. PROVIDE COMFORT II included a U.S. composite wing of reconnaissance, air superiority, and ground attack aircraft deployed in southern Turkey, plus smaller numbers of French and British aircraft. SOUTHERN WATCH has involved the 4404th Composite Wing (Provisional), with squadrons of F-15 and F-16 aircraft, plus smaller numbers of French Mirages and British Tornadoes. In addition, these operations have required specialized aircraft, including E-3, EF-111, HC-130, KC-135, RC-135, and F-16 specialized in air defense suppression.

HUMANITARIAN INTERVENTION

Humanitarian intervention is use of military force to ensure that aid reaches the intended recipients during a crisis or conflict that disrupts civil order.

Frequency

Since 1990, U.S. forces have conducted three operations that qualify as humanitarian intervention: PROVIDE COMFORT I (Northern Iraq), RESTORE HOPE (Somalia), and PROVIDE PROMISE (Bosnia-Herzegovina). At this rate, U.S. forces would conduct humanitarian intervention about every other year.

Duration

These operations took months or years and often had no satisfactory conclusion. PROVIDE COMFORT I lasted about three months (April 6–July 15, 1991), but it was followed by a no-fly operation (PROVIDE COMFORT II/ NORTHERN WATCH) that had a similar aim and still continues. The best-known humanitarian intervention, RESTORE HOPE, lasted five months (December 3, 1992–May 4, 1993) and was followed by CONTINUE HOPE, which lasted another 11 months (May 5, 1992–March 31, 1994). However, U.S. forces had a very restricted mission during the last five months of CONTINUE HOPE. After the U.S. withdrawal in March 1994, the Second United Nations Operation in Somalia (UNOSOM II) continued for another year, ostensibly trying to implement peace accords, until it withdrew under U.S. protection (UNITED SHIELD). PROVIDE PROMISE lasted over two years (July 3, 1992–October 1, 1994).

Level of Effort

Humanitarian intervention typically requires a joint task force built around the ground component, either Army or Marine forces. RESTORE HOPE included two brigades of the Army's 10th Mountain Division, as well as extensive divisional and nondivisional support, a Marine Expeditionary Brigade, a carrier battle group, an amphibious ready group, and a Maritime Prepositioning Squadron.

PEACE ACCORDS

The United States has participated in two radically different types of operations concerned with peace accords: (1) use of military forces to facilitate peace accords without coercion, usually conducted under Chapter VI of the Charter of the United Nations, and (2) use of military forces to enforce provisions of peace accords, even against the will of a party, usually conducted under Chapter VII of the Charter.

Frequency

Since 1990, U.S. forces have participated in at least six operations intended to implement or enforce peace accords: PROVIDE TRANSITION (Angola), CONTINUE HOPE (Somalia), UPHOLD DEMOCRACY/MAINTAIN DEMOCRACY (Haiti), RESTORE DEMOCRACY (Haiti), and JOINT ENDEAVOR and JOINT GUARD (Bosnia-Herzegovina). Depending on interpretation, VIGILANT SENTINEL (Kuwait) and UNITED SHIELD (Haiti) might also be included in this category. At this rate, U.S. forces would conduct such an operation yearly.

Duration

Peace accords usually entail processes that take time to conduct, such as cantonment of forces, demobilization, reconstruction, and electoral activities, implying peace operations that last months and years. PROVIDE TRANSITION, an airlift supporting U.N. peace operations in Angola, lasted over two months (August 5–October 8, 1992). UPHOLD DEMOCRACY/MAINTAIN DEMOCRACY took six months (September 19, 1994–March 30 1995) and was followed by

RESTORE DEMOCRACY, which lasted over a year (March 31, 1995–April 15, 1996). JOINT ENDEAVOR, designed to enforce the Dayton Agreements, lasted one year (December 1995–December 1996) and was immediately followed by JOINT GUARD, a smaller operation with essentially the same mission.

Level of Effort

The level of effort depends critically upon whether the operation is intended to implement or to enforce peace accords. *Peace enforcement*, as exemplified by UPHOLD DEMOCRACY/MAINTAIN DEMOCRACY and JOINT ENDEAVOR/JOINT GUARD, usually requires large joint task forces built around Army light or heavy brigades. CONTINUE HOPE is the exception that proves this rule. It was conducted with very limited forces in support of the much larger Second United Nations Operation in Somalia and incurred casualties that were considered unacceptable.

PROVIDE TRANSITION was an airlift in support of the Second United Nations Angola Verification Mission (UNAVEM II). UNAVEM II involved three C-130 aircraft from 37th Airlift Squadron and included 326 missions transporting demobilized soldiers and supplies in Angola and Zaire.

CONTINUE HOPE initially included one light infantry battalion as a Quick Reaction Force, and logistics units. During summer 1993, the United States deployed special operations forces, including a Ranger battalion, attack helicopters, and assault helicopters, in an attempt to apprehend Mohammed Farah Aideed. After October 3, 1993, when U.S. special operations forces suffered severe casualties, the United States deployed additional forces, which included a small armored task force (18 tanks and 44 infantry fighting vehicles), approximately 700 troops from 10th Mountain Division, and 4 AC-130H gunships. In addition, a carrier battle group, an amphibious ready group, and a Marine Expeditionary Unit (MEU) were available.

UPHOLD DEMOCRACY initially involved large forces, including elements of 82nd Airborne Division (to force entry if the Cedras regime refused to consent), a carrier battle group, a Marine Expeditionary Unit, and a brigade of 10th Mountain Division. As it became apparent that neither the Cedras regime nor its paramilitary supporters would offer resistance, this force was reduced to a core of Army light infantry and special forces.

During the initial phase of JOINT ENDEAVOR, U.S. land forces in Bosnia-Herzegovina were held below a ceiling of 20,000 troops, approximately one-third of NATO's deployed strength. The U.S. initially deployed two heavy brigades, each containing one armor battalion and one mechanized infantry battalion. These forces were supported by a composite air wing in Aviano and, at times, by a carrier battle group and an amphibious ready group in the Adriatic or Mediterranean. After the parties observed the cease-fire, withdrew from the zone of separation, and placed their heavy weapons at collection points, the United States withdrew its armor battalions and deployed additional military police companies. U.S. and other NATO forces conducted JOINT GUARD at approximately half of what their strength was during JOINT ENDEAVOR.

SUMMARY OF OPERATIONS, 1990–1996

Since the end of the Cold War, the United States has conducted operations more often and in greater force. With few exceptions, U.S. forces have accomplished their missions successfully.

Summary and Timeline

At the "high end," i.e., while conducting humanitarian intervention and peace enforcement, operations have been nearly continuous since 1992. Deployed combat strength has also risen, culminating in the two well-supported Army heavy brigades that initially entered Bosnia-Herzegovina (see Figure 2.1).

Land- and sea-based air forces have conducted continuous operations simultaneously in the former Yugoslavia and in Iraq to enforce no-fly zones, to conduct ground attacks, and to support land forces. The supported forces have included a multinational force in Northern Iraq during PROVIDE COMFORT I, the United Nations Protection Force in Bosnia-Herzegovina during DENY FLIGHT, the NATO-led Implementation Force during JOINT ENDEAVOR, and the NATO-led Stabilization Force during JOINT GUARD. At the same time, U.S. forces continued to conduct battalion-sized peacekeeping

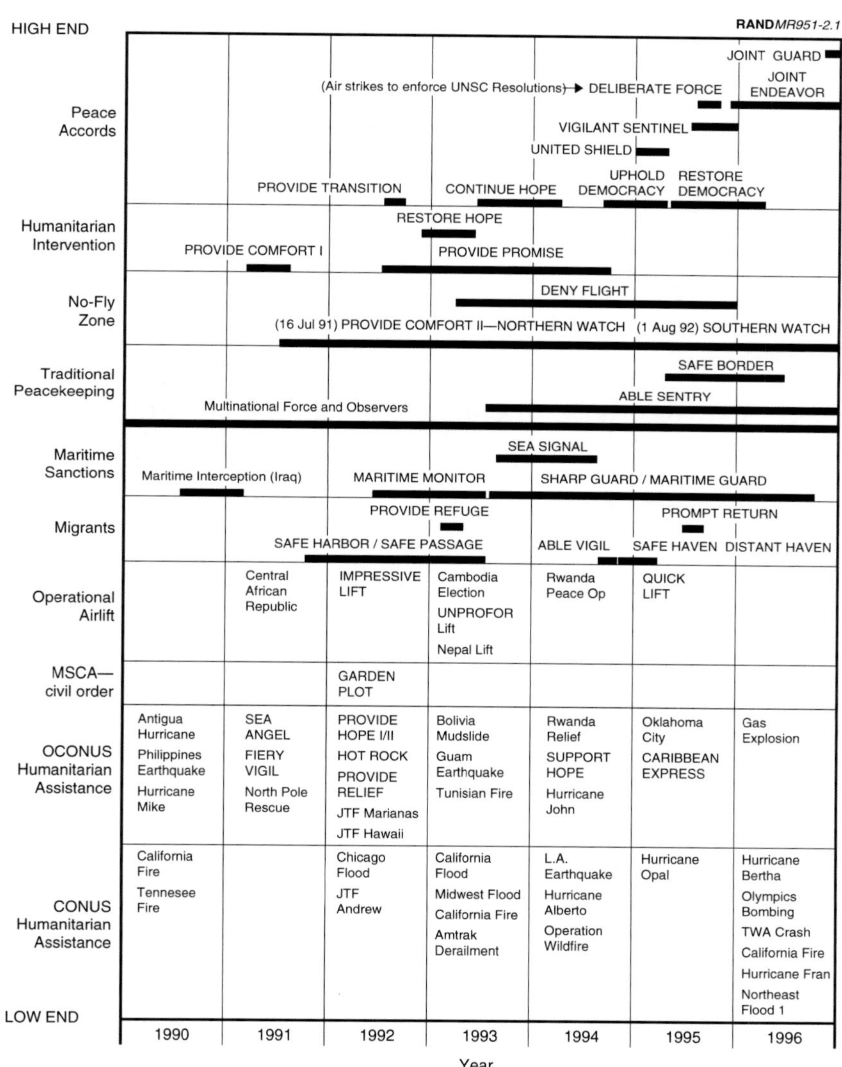

NOTE: CONUS = continental United States; JTF = Joint Task Force; MSCA = military support to civil authorities; OCONUS = outside the continental United States; UNPROFOR = U.N. Protection Force; UNSC = United Nations Security Council.

Figure 2.1—Summary of Operations Conducted 1990–1996

operations in Macedonia (ABLE SENTRY) and the Sinai (Multinational Force and Observers). Despite their minuscule size, peacekeeping operations have large cumulative effects because they are protracted indefinitely.

Success and Failure

How successfully have U.S. forces conducted these operations? Have there been failures that suggest shortfalls in capability? Over the seven years of interest here, U.S. forces have been consistently successful when deployed in sufficient strength. No significant shortfalls have been apparent. Indeed, it would be surprising if forces developed to withstand an onslaught of the former Soviet Union's vast forces and to prevail in major theater warfare were not able to secure humanitarian aid or to enforce peace agreements among small factions and minor regional powers. However, operational/tactical outcomes may be very different from strategic outcomes. Even the most successful peace operation may not yield the desired strategic result over the long term. For example, UPHOLD DEMOCRACY, the major U.S. operation in Haiti, was immensely successful, yet Haiti might revert to ill-governance and oppression. UPHOLD DEMOCRACY gave Haiti a chance for decent government, not a guarantee (see Figure 2.2).

At the tactical and operational levels, U.S. forces have enjoyed almost perfect success. There have been blemishes, such as the destruction of two Army helicopters by Air Force pilots during PROVIDE COMFORT II (now NORTHERN WATCH), but no failures except CONTINUE HOPE. Moreover, this apparent exception actually proves the rule: U.S. forces failed in Somalia not because they lacked capabilities but because some capabilities were withheld for political reasons.[2] Had CONTINUE HOPE been conducted with forces at the

---

[2]Unduly complicated command arrangements and inadequate support from United Nations forces contributed to the losses suffered during CONTINUE HOPE, but the fundamental cause was an inadequate U.S. force structure. Maj. Gen. Thomas Montgomery, commanding U.S. troops in Somalia (less special operations forces) requested armored and mechanized forces, but his request was denied for political reasons. As a result, U.S. commanders in Somalia continued to conduct risky special operations that eventually suffered high casualties. Moreover, when U.S. special

RANDMR951-2.2

Operation	Tactical/Operational Outcome	Strategic Successes/Failures
PROVIDE COMFORT I	Provided relief supplies to Kurdish refugees, established refugee camps, compelled withdrawal of Iraqi forces, protected return of refugees.	Prevented many Kurdish deaths to starvation and exposure; temporarily curtailed Saddam Hussein's power in northern Iraq.
PROVIDE COMFORT II	Enforced no-fly zone over northern Iraq; inadvertently destroyed two U.S. helicopters.	Failed to protect Kurds: in September 1996, Iraqi ground forces entered the area, compelling an evacuation.
RESTORE HOPE	Secured ports, roads, and distribution points; dismantled checkpoints, confiscated unauthorized weapons; ensured receipt of humanitarian aid.	Ended mass starvation in Somalia, but did not end factional warfare, especially in the Mogadishu area.
CONTINUE HOPE	Provided Quick Reaction Force and logistics support to UNOSOM II; failed to apprehend Mohammed Farah Aideed.	Failed to make UNOSOM II viable; Somalia remained plagued by factional warfare; U.S. administration and U.N. were discredited.
UPHOLD DEMOCRACY	Restored Aristide government; provided security; monitored police; seized weapons.	Gave Haiti a chance for decent government; ended source of illegal migration to U.S.
SHARP GUARD	Interdicted weapons shipments by sea until U.S. unilaterally withdrew from the operation.	Worked to advantage of better-armed Bosnian Serbs; parties were supplied by air and land.
DENY FLIGHT	Prevented flight of fixed-wing aircraft; provided close air support to UNPROFOR.	Neutralized Bosnian Serb air forces, but UNPROFOR was too weak to benefit from close air support.
PROVIDE PROMISE	Airlifted humanitarian aid to Sarajevo; airdropped aid in Muslim-held enclaves.	Helped Muslims withstand siege of Sarajevo and retain enclaves for a time.
DELIBERATE FORCE	Destroyed Bosnian Serb military installations with little collateral damage.	Made Bosnian Serbs respect "safe areas"; helped lead to Dayton Agreements.
JOINT ENDEAVOR	Enforced cease-fire, withdrawal from zone of separation, cantonment of heavy weapons.	Brought peace to Bosnia; restored NATO's prestige; affirmed U.S. leadership in Europe.

NOTE: UNOSOM II = Second United Nations Operation in Somalia; UNPROFOR = U.N. Protection Force.

Figure 2.2—Evaluation of Selected Operations

same level as RESTORE HOPE, it would certainly have succeeded. Even at a reduced level, CONTINUE HOPE might have succeeded if the armor and mechanized infantry had been received as requested.

---

operations forces were pinned down by fire, the U.S. Quick Reaction Force could not reach them until U.N. forces eventually provided tanks and infantry fighting vehicles.

In the following chapter, we use the historical record of past operations to construct vignettes, which are essentially archetypes that abstract from actual operations by developing simplified patterns.

Chapter Three

VIGNETTES

We next constructed a series of vignettes. The purpose of these vignettes is to give a clear picture of requirements by stripping away the quirkiness of historical operations. For example, during an historical operation, several units may have contributed personnel and equipment to form a unit-equivalent; the corresponding vignette shows just that *type* of unit, not a collection of fragments. As a result, the vignettes provide a simple, readily intelligible view of force requirements.

For each vignette, we developed a mission statement, phases with implied tasks, and lists of units for each phase. We developed detailed force lists at the following levels of aggregation:

- Air Force: squadron level
- Navy: ship class
- Marine Corps: battalion, squadron, separate company
- Army: battalion, separate company, detachment, and team.

We entered typical personnel strengths at full authorization for each unit and computed total personnel strength for each service during each phase.

Each vignette ends with return of forces to either their home station or their normal operating area. However, following protracted deployment, forces require a period to reconstitute and retrain. *Reconstitution* includes repair and replacement of equipment. *Retraining* includes refreshment of skills that declined during the operations, which restrict opportunities to train. Maneuver forces typically have to refresh such skills as marksmanship, gunnery, and combined-arms tactics. Fighter pilots must refresh their skills in air-to-air engagement, which atrophy during the monotonous routine of peace operations, such as enforcing no-fly zones. Reconstitution and retraining may require months to complete. During this time, some units would not be expected to deploy again, unless in a great emergency or war.

CONUS/OCONUS HUMANITARIAN ASSISTANCE (MEDIUM DISASTER)

This vignette depicts an operation comparable to those conducted following Hurricanes Iniki, Omar, and Mike, and Typhoon Offa.

Mission

Provide military support to civil authorities following a hurricane that has caused considerable damage within the continental United States, Hawaii, Alaska, or U.S. territories. The environment is co-operative, and the threat is limited to looting and lawlessness during the initial phase of operations.

Phases

Phase One (Week 1): Establish command and control, usually through a small joint task force. Conduct liaison with civil authorities. Deploy forces into an affected area. Conduct search-and-rescue operations. Evacuate civilians. Assist civil law enforcement. Provide emergency communications and electrical power. Provide potable water, food, emergency medical care, and shelter to survivors. Clear debris and make initial repairs to infrastructure.

Forces are likely to include the following:

- Air Force: one or more airlift squadrons, aeromedical evacuation squadron, aerial port squadron
- Navy: one or more amphibious ships or Coast Guard cutters, deep submergence craft, rescue and salvage ship

- Marine Corps: elements drawn from infantry, engineer, communications, logistics, medium- and heavy-helicopter units
- Army: elements drawn from infantry, military police, signal, engineer, medical, preventive medicine, water-purification, general-purpose helicopter, medium helicopter, bridging, light–medium truck, civil affairs, and public affairs units.

Phase Two (Weeks 2 to 3): Continue clearing debris and making initial repairs to infrastructure. Assist in restoring harbors and raising sunken vessels if required. Return forces to home stations. Forces are similar to those in Phase One, declining as civil authorities assume responsibility.

CONUS HUMANITARIAN ASSISTANCE (LARGE DISASTER)

This vignette depicts an operation comparable to JTF Andrew (assistance following Hurricane Andrew, which caused devastation in Florida and Louisiana during 1992).

Mission

Provide military support to civil authorities following a major natural disaster that has caused extensive damage within the continental United States. The environment is cooperative, and the threat is limited to looting and lawlessness during the initial phase of operations.

Phases

Phase One (Week 1): Conduct liaison with civil authorities. Establish a joint task force to control and integrate military support. Deploy forces into an area suffering extensive damage. Conduct search-and-rescue operations. Support civil law enforcement in maintaining public order. Evacuate civilians. Provide emergency communications and electrical power. Provide potable water, food, emergency medical care, and shelter for survivors. Restore lines of communication.

Forces are likely to include the following:

- Air Force: 2–3 airlift squadrons, aeromedical evacuation squadron
- Navy: Coast Guard cutters, deep submergence craft, rescue and salvage ship
- Marine Corps: elements drawn from infantry, engineer, communications, logistics, medium- and heavy-helicopter units
- Army: elements drawn from corps and divisional headquarters, 3 light or airborne infantry battalions, military police brigade, signal battalion, engineer brigade, medical group, 2–4 water-purification detachments, aviation brigade, 2–3 light–medium truck companies, division support command, movement control detachment, cargo transfer company, forward support battalion, 1–2 maintenance companies, petroleum-supply battalion, civil affairs element, public affairs element.

Phase Two (Weeks 2 through 6): Continue to provide potable water, food, emergency medical care, and shelter to survivors as required. Clear debris and make initial repairs to infrastructure, especially to sanitation systems. Assist in restoring harbor facilities and raising sunken vessels if required. Transition to operations conducted solely by civilian authorities. Return forces to home stations. Forces are similar to those in Phase One. Force levels peak during Phase Two as units complete their deployment, then decline as civil authorities assume responsibility.

OCONUS HUMANITARIAN ASSISTANCE (LITTORAL OPERATION)

This vignette depicts an operation comparable to SEA ANGEL, conducted in Bangladesh during 1991, but it employs a Marine Expeditionary Unit (built around a single Marine infantry battalion) rather than a Marine Expeditionary Brigade (built around multiple Marine infantry battalions).

Mission

Provide military support to foreign civil authorities following a major natural disaster occurring on a littoral. The environment is cooperative, and the threat may be limited to potential looters and harassment by small groups of dissidents.

Phases

Phase One (Week 1): Conduct liaison with foreign civil authorities and U.S. authorities in-country. Establish a joint task force to control U.S. forces. Deploy forces into an area of operations. Conduct search-and-rescue operations. Evacuate civilians. Survey damage and estimate requirements for assistance and repair. Develop seaports, airports, overland transportation, and storage facilities as required to support operations. Provide potable water, emergency rations, immediate medical care, and shelter to an affected population. Support government agencies and nongovernmental organizations (NGOs) engaged in disaster relief. Assist in emergency repairs to a country's infrastructure.

Forces are likely to include the following:

- Air Force: a composite airlift wing, aerial port squadron, aeromedical evacuation squadron, air support operations squadron, special operations squadron (MC-130), special tactics squadron
- Navy: amphibious ready group, including two amphibious assault ships and one dock landing ship
- Marine Corps: one Marine Expeditionary Unit
- Army: special forces element, engineer company, general-support helicopter company.

Phase Two (Weeks 2 through 5): Continue to provide potable water, emergency rations, immediate medical care, and shelter as necessary. Support government agencies and nongovernmental organizations engaged in disaster relief. Prepare civil authorities and NGOs for departure of U.S. forces. Redeploy forces to home stations. Forces are similar to those in Phase One.

PEACEKEEPING (PREVENTIVE DEPLOYMENT)

This vignette depicts peacekeeping with an implied commitment to respond if violations occur. Examples are ABLE SENTRY (Macedonia) and the United Nations Iraq-Kuwait Observation Mission (UNIKOM), both currently in progress.[1] An example of peacekeeping without this implied commitment is the Multinational Force and Observers in the Sinai.

Mission

Under authority of a resolution of the United Nations Security Council, monitor and report on activity in a border area. Be prepared to secure a border area against incursion if required. The environment is initially permissive, but potentially hostile. The threat ranges from border violations by armed individuals to large-scale incursions by heavily armed conventional forces. However, the peacekeeping force is not expected to oppose large-scale incursions.

Phases

Phase One (Months 1 to *D*): Establish an international or multilateral command entity to control operations. Conduct liaison with a host country and neighboring countries as appropriate. Conduct space, aerial, naval, and ground surveillance as appropriate. Establish observation points and conduct patrols. Monitor and report on developments that could undermine stability or threaten the territorial sovereignty of a host country. Establish a demilitarized zone in a border area and control entry to this zone, if such a zone is declared. Forces are likely to include the following:

- one infantry battalion with augmented staff
- logistics support
- aviation element (UH-60).

---

[1]The small, light forces deployed in ABLE SENTRY and UNIKOM could not respond effectively to serious violations. If violations seemed imminent or actually occurred, these forces would have to be augmented or replaced by larger, more heavily armed forces that could respond, as occurred on several occasions in Kuwait.

Phase Two (Months *D* to *D+X*): If events occur that threaten stability (Month *D*), deploy additional forces to demonstrate resolve and secure a border area against incursion. Establish air supremacy and sea control.

Forces are likely to include the following:

- Air Force: composite wing (F-15, F-16, A-10), reconnaissance elements (U-2, RC-135), electronic combat squadron (EC-130), special operations squadron (MC-130), special operations squadron (AC-130), airborne control squadron (E-3), composite airlift wing, 2 air-refueling squadrons (KC-135, KC-10), airlift control flight, air intelligence squadron, aeromedical evacuation squadron, aerial port squadron, maintenance squadron, munitions squadron
- Navy: carrier battle group, amphibious ready group
- Marine Corps: Marine Expeditionary Unit, which may go ashore or be held in readiness
- Army: heavy brigade (armor and mechanized infantry), field artillery battalion, medium-helicopter battalion, attack-helicopter battalion, aviation maintenance battalion, signal element, civil affairs element, public affairs detachment.

Phase Three (Months *D+X* through *N*): Return to the mission accomplished during Phase One. Forces might return to the level of those in Phase One or be reinforced.

NO-FLY ZONE AND STRIKE

This vignette depicts an operation to enforce a no-fly zone and to conduct air strikes as required. The operation is comparable to DENY FLIGHT/DELIBERATE FORCE/DECISIVE EDGE (Bosnia-Herzegovina), PROVIDE COMFORT II/NORTHERN WATCH (Northern Iraq), and SOUTHERN WATCH (Southern Iraq).

Mission

As part of an international force, enforce a no-fly zone. On command, conduct ground attacks to coerce parties or to punish recalcitrant behavior. The environment is potentially hostile. The threat includes substantial numbers of heavy machine guns, cannons, hand-held infrared-seeking missiles, fixed and mobile air-defense missiles, and fighter aircraft.

Phases

Phase One (Months 1–*D*): Establish joint and combined command relationships and command entities to control the operation. Establish control over authorized air traffic within the no-fly zone. Establish reliable means of friend-or-foe identification. Deploy land-based and carrier-based air forces into the area of operations. Protect the deployed forces. Maintain continuous, near-real-time surveillance of the no-fly zone and related airspace. Maintain surveillance of air bases and airfields in the area of operations. Maintain combat air forces on patrol or strip alert sufficient to ensure local air superiority within the no-fly zone. Air-refuel those aircraft conducting airborne warning and control, reconnaissance, and combat air patrol. Intercept, warn, or destroy, if necessary, aircraft that violate the zone. Provide search and rescue. Prepare to conduct punitive strikes on command.

Forces are likely to include the following:

- Air Force: composite wing (F-15, F-16, O/A-10), reconnaissance elements (U-2, RC-135), electronic combat squadron (EC-130), special operations squadron (MC-130), special operations squadron (MH-53), special operations squadron (AC-130), airborne control squadron (E-3), air-refueling squadron (KC-135), airlift control flight, air intelligence squadron, maintenance squadron, munitions squadron
- Navy: carrier battle group and amphibious ready group, both available periodically
- Marine Corps: fighter attack squadron (FA-18), elements of a Marine Expeditionary Unit—Special Operations Capable (MEU [SOC]) to conduct search and rescue, available periodically
- Army: special operations forces to conduct search and rescue.

Phase Two (Month *D*): Reinforce the deployed air forces. Develop an approved target list appropriate to the intended purpose. Develop a concept of operations and designate units to accomplish the implied tasks. Generate a combined air tasking order for those U.S. and allied air forces that will conduct the operation. Provide real-time control of ingressing and egressing aircraft. Suppress air defenses. Attack targets with appropriate munitions mixes. Take measures to minimize collateral damage. Assess damage to targets and re-engage as required. Provide search and rescue.

Forces are likely to include the following:

- Air Force: same as Phase One, but reinforced by additional strike and electronic combat aircraft
- Navy: carrier battle group and amphibious ready group on-station
- Marine Corps: fighter attack squadron (FA-18), elements of a MEU (SOC) to conduct search and rescue
- Army: special operations forces to conduct search and rescue.

Phase Three (Months *D–X*): Conduct air-denial operations as during Phase One and be prepared to conduct punitive strikes. Forces are the same as those in Phase One.

HUMANITARIAN INTERVENTION (INLAND)

This vignette depicts a small humanitarian intervention hundreds of kilometers inland, such as PROVIDE COMFORT I (Northern Iraq).

Mission

Within an international or multinational effort, secure and provide humanitarian assistance to a population suffering as a result of conflict. The environment is initially permissive, but potentially hostile. The threat ranges from petty harassment to large-scale opposition by conventional forces.

Phases

Phase One (Month 1): Establish international or multilateral command and control arrangements. Conduct liaison with host countries. Gain and maintain air supremacy and sea control in an area of operations. Conduct surveillance and reconnaissance to support an estimate of need and to provide warning of interference. Develop seaports, airports, overland transport, and storage facilities as required to support operations. Deploy forces into an area of operations. Protect deployed forces. Secure delivery of humanitarian aid to the intended recipients. Deter and prevent armed obstruction of relief operations. Establish temporary camps for displaced civilians. Provide potable water, emergency rations, immediate medical care, and shelter. Establish a civilian-military operations center to coordinate efforts. Support nongovernmental organizations engaged in humanitarian assistance. If required, evacuate endangered civilians employed by NGOs.

Forces are likely to include the following:

- Air Force: fighter squadron, electronic combat squadron (EC-130), reconnaissance squadron, special operations squadron (MC-130), special operations squadron (AC-130), composite airlift wing, air-refueling squadron, airlift control flight, air intelligence squadron, aeromedical evacuation squadron, aerial port squadron, maintenance squadron, munitions squadron
- Navy: carrier battle group, amphibious ready group
- Marine Corps: Marine Expeditionary Unit
- Army: 2 special forces battalions, medium-helicopter battalion, aviation maintenance battalion, signal element, civil affairs element, public affairs team.

Phase Two (Months 2 through 4): Continue to deter and prevent armed obstruction of relief operations. Provide potable water, emergency rations, immediate medical care, and shelter as required. Support NGOs engaged in humanitarian assistance. Return civilians to their homes. Prepare civil authorities and NGOs for departure of forces. Redeploy forces to home stations.

Forces are similar to those in Phase One. The carrier battle group would operate elsewhere in the region but become available again if a crisis occurred.

HUMANITARIAN INTERVENTION AND PEACE ACCORD ENFORCEMENT

This vignette depicts a large-scale humanitarian intervention followed by a coercive (Chapter VII) operation to enforce a peace accord. Examples are RESTORE HOPE and the Second United Nations Operation in Somalia (UNOSOM II).

Mission

Intervene to secure delivery of humanitarian assistance during a conflict. Subsequently, support an international or multinational operation to enforce a peace accord. The environment is initially permissive, but potentially hostile. The threat may include looters, bandits, paramilitary forces, and lightly armed militias.

Phases

Phase One (Months 1–4): Establish a joint task force to control U.S. forces. Conduct reconnaissance in an area of operations (AO). Gain air supremacy and sea control. Secure major airports, seaports, other key installations, and lines of communication. Deploy U.S. forces. Gain freedom of movement and show overwhelming force to warring factions. Dismantle unauthorized checkpoints and suppress banditry. Ensure free passage of humanitarian assistance. Secure personnel and equipment of NGOs. Provide logistics support to NGOs. Conduct disarmament as necessary to establish a secure environment. Repair key infrastructure, including roads, bridges, seaport facilities, and airport facilities, to support operations. Mark and remove land mines as required.

Forces are likely to include the following:

- Air Force: reconnaissance elements (U-2, RC-135), electronic combat squadron (EC-130), special operations squadron (MC-130), special operations squadron (AC-130), 2 composite airlift wings, 2 air-refueling squadrons (KC-135, KC-10), airlift control flight, air intelligence squadron, aeromedical evacuation squadron, aerial port squadron, engineer squadron (RED HORSE),[2] security police squadron

- Navy: carrier battle group, amphibious ready group
- Marine Corps: Marine Expeditionary Brigade built around 2 infantry battalions, including attack squadron, medium-helicopter squadron, heavy-helicopter squadron, aviation logistics squadron
- Army: task force built around a light-infantry brigade, including 2 special forces battalions, tank battalion, mechanized infantry battalion, 2 military police companies, composite aviation brigade, psychological operations battalion, and tailored elements of a division support command.

Phase Two (Months 5–12): Support an international or multilateral operation to enforce peace accords, typically including separation of forces, cantonment of weapons, disarmament and demobilization, repair of infrastructure, reconstitution of civil authority, and support of electoral activities. Maintain freedom of movement for U.S. and other nations' forces. Provide a quick-reaction force to coerce recalcitrant factions if necessary. Provide logistics support to multinational forces and NGOs.

Forces are likely to include the following:

- Air Force: as during Phase One, but with much-reduced requirement for airlift and air-refueling
- Navy: carrier battle group and amphibious ready group depart the area
- Marine Corps: Marine Expeditionary Brigade redeploys
- Army: task force built around a composite brigade, including 2 light-infantry battalions, tank battalion, mechanized infantry battalion, special forces battalion, Ranger battalion, 2 military

---

[2]RED HORSE stands for Rapid, Engineer-Deployable, Heavy Operational Repair Squadron, Engineer. It is described in more detail in Chapter Four.

police companies, composite aviation brigade, and tailored elements of a division support command.[3]

Phase Three (Months 13–24): Participate in an international or multinational operation to monitor a continuing peace process.

Forces are likely to include the following:

- Air Force: elements of an airlift squadron and aerial port squadron
- Navy: none
- Marine Corps: none
- Army: task force built around an infantry battalion, including military police company, general-support helicopter battalion, preventive medicine detachments, and ordnance-disposal team.

PEACE ACCORD ENFORCEMENT (LEGITIMATE GOVERNMENT)

This vignette depicts an operation, comparable to UPHOLD DEMOCRACY/MAINTAIN DEMOCRACY (Haiti), to enforce an accord that ensures restoration of a legitimate government.

Mission

Use force as necessary to ensure restoration of a legitimate government. Subsequently, participate in a multinational operation to support the newly restored government. The environment is initially

---

[3]This force is much larger than the initial force during the historical example (UNOSOM II). The initial force was built around one light-infantry battalion, designated as a quick-reaction force. Subsequently, the United States deployed special operations forces under separate command arrangements. After the special operations forces lost 18 men during a firefight on October 3, 1993, the United States deployed two heavy battalions and one light battalion to Somalia and placed a carrier battle group and amphibious ready group offshore. At the same time, the United States reduced the mission to force protection, security of humanitarian aid, and presence. The force structure depicted in our vignette is based on a judgment that the Army forces sent after October would have been adequate to perform the original mission.

permissive, but may become hostile. The threat includes criminal elements, paramilitary groups, and small, lightly armed regular forces during the initial phase of operations.

Phases

Phase One (Month 1): As time permits, plan the operation and exercise key parts of the plan. Gain air supremacy and sea control in an area of operations. Be prepared to force entry if a *de facto* regime offers resistance. Deploy forces into an area of operations, either administratively or tactically as required by the situation. Show an overwhelming force. Secure key areas, especially urban areas, and lines of communication. Establish checkpoints and conduct patrols. Maintain civil authority. Ensure restoration of a legitimate government. Provide security to government officials.

Forces are likely to include the following:

- Air Force: fighter squadron (F-15/F-16), reconnaissance elements (U-2, RC-135), electronic combat squadron (EC-130), special operations squadron (MC-130), special operations squadron (AC-130), airborne control squadron (E-3), composite airlift wing, 2 air-refueling squadrons (KC-135, KC-10), airlift control flight, air intelligence squadron, aeromedical evacuation squadron, aerial port squadron
- Navy: carrier battle group, amphibious ready group
- Marine Corps: Marine Expeditionary Unit
- Army: task force built around 2 light-infantry brigades including a special forces group, a composite aviation brigade, a military police battalion, a medical group, and a tailored division support command.

Phase Two (Months 2–6): Maintain a secure environment for a legitimate government. Confiscate unauthorized weapons, assist in reconstitution of civil authority, and repair infrastructure. Conduct civic action programs. Transition to a multinational operation. Redeploy most U.S. forces to home stations.

Forces are likely to include the following:

- Air Force: as during Phase One, except that fighter squadron may not be required
- Navy: carrier battle group and amphibious ready group depart
- Marine Corps: Marine Expeditionary Unit redeploys
- Army: as during Phase One, except that one infantry brigade redeploys and gradual drawdown begins.

Phase Three (Months 7–18): Assist a multinational operation supporting a legitimate government. Provide a quick-reaction force. Assist in securing and monitoring electoral activities. Provide logistics support to multinational forces, international organizations, and nongovernmental organizations.

Forces are likely to include the following:

- Army: command element, infantry battalion (light or motorized), special forces company, military police company, general-support helicopter battalion (UH-60), support element.

PEACE ACCORD ENFORCEMENT (AGREEMENT AMONG PARTIES)

This vignette depicts an operation to enforce a peace accord concluded among parties, such as JOINT ENDEAVOR/JOINT GUARD (Bosnia-Herzegovina, December 1995 to present).

Mission

As part of an international force, enforce a peace agreement. Show overwhelming force and be prepared to coerce recalcitrant parties. The environment is initially permissive, but potentially hostile. The threat ranges from small-scale attacks with unconventional forces to large-scale attacks with heavily armed conventional forces conducted by one of the parties to the peace accord.

Phases

Phase One (Months 1–4): As time permits, plan the operation and exercise key parts of the plan. Establish command relationships and command entities to control the operation. Establish relations with parties to the conflict, other allied and friendly forces, international organizations, government agencies, and nongovernmental organizations. Deploy forces into the area of operations. Protect the force, maintain freedom of movement, and show overwhelming force to the parties. Maintain air supremacy in the area of operations and secure lines of communication. Enforce military provisions of a peace accord, typically including separation of forces, destruction of obstacles and fortifications, cantonment of heavy weapons, disarmament, and demobilization. Repair infrastructure as required to support operations. Mark and remove land mines as required. Respond to serious violations of human rights if they occur.

Forces are likely to include the following:

- Air Force: composite wing (F-15, F-16, A-10), reconnaissance elements (U-2, RC-135), electronic combat squadron (EC-130), special operations squadron (MC-130), special operations squadron (AC-130), 2 composite airlift wings, 2 air-refueling squadrons (KC-135, KC-10), airlift control squadron, air operations squadron, air intelligence squadron, aeromedical evacuation squadron, medical operations squadron, 2–3 aerial port squadrons, engineer squadron (RED HORSE), security police squadron
- Navy: carrier battle group, amphibious ready group
- Marine Corps: Marine Expeditionary Unit
- Army: task force built around 2 heavy brigades, including special forces battalion, military police battalion, composite aviation brigade, engineer brigade, civil affairs battalion, psychological operations battalion, and tailored elements of corps/division support command.

Phase Two (Months 5–12): Ensure continued compliance with military provisions of the peace accord. Maintain freedom of movement and be prepared to show overwhelming force. Within available resources, help to repair infrastructure, reconstitute civil authority, and support electoral activities. Respond to serious violations of human rights if they occur. Monitor clearance of minefields and provide technical assistance.

Forces are likely to include the following:

- Air Force: same as in Phase One, but with reduced requirement for airlift and airport operations
- Navy: carrier battle group and amphibious ready group remain in the region, but may respond to other tasking
- Marine Corps: Marine Expeditionary Unit remains in the region, but may respond to other tasking
- Army: same as in Phase One, but 2 heavy battalions are replaced by 2 additional military police battalions.

Phase Three (Months 13–24): Maintain a stable environment conducive to a peace process. Maintain freedom of movement and be prepared to show overwhelming force, using reinforcements if necessary. Within available resources, assist civilian agencies engaged in reconstruction and reconstitution of civil government.

Forces are likely to include the following:

- Air Force: same as in Phase Two, but with further reduction in requirement for airlift and airport operations
- Navy: carrier battle group and amphibious ready group remain in the region, but may respond to other tasking
- Marine Corps: Marine Expeditionary Unit remains in the region, but may respond to other tasking
- Army: task force built around one composite brigade.

Phase Four (Months 25–30): Monitor continuance of a peace process and prepare for withdrawal. Redeploy forces to home stations. Remaining forces are likely to include the following:

- Air Force: small composite fighter wing
- Navy: carrier battle group and amphibious ready group remain in the region, but may respond to other tasking

- Marine Corps: Marine Expeditionary Unit remains in the region, but may respond to other tasking
- Army: task force built around one mechanized infantry battalion.[4]

SUMMARY

Most of the operations depicted in these vignettes could involve combat, and potential opposition is the primary consideration in selecting the size and composition of U.S. forces. Moreover, high-end operations require overwhelming forces to deter opposition that could upset peace accords and turn U.S. public opinion against such operations. Figure 3.1 summarizes the factors that determine force requirements.

Only the first three vignettes, all centering on humanitarian assistance, do not imply combat operations. Factors that determine the configuration of U.S. forces include the geographic characteristics of the area of operations, distance from CONUS and support bases, the nature and extent of devastation, and the needs of the civilian population that is affected by the disaster. Following an unusually severe disaster, there may also be a requirement to assist civil authorities in restoring public order for a short period of time.

Preventive deployment initially requires small forces, but potentially demands large reinforcements. The peacekeeping force is not expected to enter combat, other than in self-defense. Therefore, potential opposition is irrelevant to the peacekeeping force, because the force is designed merely to observe certain activities within an area of operations. But if violations occur, as they did in Kuwait, the United States is obligated to deploy additional forces that are prepared to fight. *Potential opposition* is the primary factor in determining size and composition of reinforcements.

---

[4]This force is not derived from historical data, because the operation used as the example (JOINT ENDEAVOR/JOINT GUARD) is currently in Phase Three. We have extrapolated from previous drawndowns to arrive at a force of this size for a projected fourth phase. Alternatively, the United States might elect to maintain a brigade-sized force in Bosnia up to the end of the mission.

RANDMR951-3.1

<table>
<tr><th>Vignette</th><th>Combat?</th><th>Determination of Force Requirements</th></tr>
<tr><td>CONUS/OCONUS Humanitarian Assistance (Medium)</td><td>no</td><td rowspan="3">Area of operations, extent of devastation, and needs of the affected population.</td></tr>
<tr><td>CONUS Humanitarian Assistance (Large Disaster)</td><td>no</td></tr>
<tr><td>OCONUS Humanitarian Assistance (Littoral Operation)</td><td>no</td></tr>
<tr><td>Peacekeeping (Preventive Deployment)</td><td>yes</td><td>1. Area and activities to be observed.
2. If violations: potential opposition.</td></tr>
<tr><td>No-Fly Zone and Strike</td><td>yes</td><td>Area of operations, target sets to be attacked, and potential opposition.</td></tr>
<tr><td>Humanitarian Intervention (Inland)</td><td>yes</td><td rowspan="2">Area of operations, available infrastructure, needs of affected population, and potential opposition (to deter opposition, force should appear overwhelming).</td></tr>
<tr><td>Humanitarian Intervention and Peace Accord Enforcement</td><td>yes</td></tr>
<tr><td>Peace Accord Enforcement (Legitimate Government)</td><td>yes</td><td rowspan="2">Area of operations, available infrastructure, and potential opposition (to deter opposition, force should appear overwhelming).</td></tr>
<tr><td>Peace Accord Enforcement (Agreement Among Parties)</td><td>yes</td></tr>
</table>

NOTE: CONUS = continental United States; OCONUS = outside the continental United States.

Figure 3.1—Factors Determining Force Requirements

All other vignettes imply that U.S. forces must be prepared to conduct combat operations under specified conditions during at least some phase of the operation. To enforce no-fly zones, U.S. forces must be configured to keep the area of operations under surveillance, destroy the designated target sets if necessary, and defeat potential opposition, usually including both air and air-defense forces. Humanitarian intervention requires forces that are configured to the area of operations, those needs of the affected population that cannot be met by other agencies, and potential opposition, which can range from individual bandits to large, conventionally armed forces—for example, Iraqi forces during PROVIDE COMFORT I. When U.S. forces are expected to enforce peace accords, they must be configured to defeat any party that might renege on its agreements. Moreover, they must appear overwhelming from the perspective of potential opponents, implying an immediate show of force that potential opponents perceive as crushing superiority.

Chapter Four

IMPLICATIONS

Requirements for military forces are sensitive to the level of future operations, which is highly uncertain. Current forces can meet peak demands, even at an increased level of operations. But they may encounter serious difficulties in the process. First, forces committed to these operations might be unavailable or not immediately available for theater warfare. Second, demands might fall very unevenly on the armed services and upon types of units within the services. To obtain certain specialized skills and to sustain rotation of units during protracted operations, the Army must access its Reserve and National Guard. The Air Force faces basing challenges and suffers highly uneven stress on aircraft and crews, contributing to a looming retention problem in personnel.

LEVEL OF FUTURE OPERATIONS

The effects and implications of future operations depend heavily on the level of high-end operations in the future. If the level diminishes, then most of the effects are past and the implications may be insignificant. But if the level of high-end operations remains constant or even increases, then the effects will mount and the implications may be profound. What level is likely?

Little uncertainty surrounds the usual run of natural disasters in the United States. Unpredictable as individual events, they exhibit a regular pattern in the aggregate over time. Hurricanes, typhoons, floods, and forest fires recur seasonally in roughly the same areas of the United States and its territories. Response to such disasters is also predictable: U.S. leaders are compelled by sympathy and political pressure to respond quickly and effectively, employing military forces as required. But great uncertainty surrounds the occurrence of a major catastrophe that could occur at any time with little or no warning.

Little uncertainty surrounds U.S. participation in traditional peacekeeping. Many other countries have the required military capabilities, but the United States participates on an exceptional basis for political reasons. It currently contributes one infantry battalion each to two such operations, a tempo that might continue indefinitely.

Great uncertainty surrounds the high-end peace operations, categorized as "humanitarian intervention" and "peace enforcement." Conflict that might prompt consideration of such operations and make them appear feasible is highly unpredictable. Even greater uncertainty surrounds U.S. responses. U.S. leaders have considerable latitude in deciding whether to become involved and at what level of effort. U.S. geopolitical interests, however defined, shed only a flickering light on the subject: The United States entered into Somalia, although it had no discernible geopolitical interest, yet held itself aloof for a long time from Bosnia, although its NATO allies were heavily involved and floundering for lack of U.S. leadership.

In the following subsections, we look at three possible levels for future operations: diminished, constant, and increased.

Diminished Level of Operations

Over the coming five years, high-end operations might diminish, for several reasons. Conflicts might not occur where the United States had sufficient geopolitical interest to justify becoming involved. Or such conflicts might occur, but not appear amenable to peace operations. They might appear intractable if the parties refused to consider peace accords or if military operations were unconventional, obviating the usual techniques of peace operations. In addition, the United States might be unable to obtain affirmative votes in the Security Council, and the U.N. might be crippled by a financial crisis precipitated by expensive peace operations and unpaid assessments. Congressional leadership might block U.S. participation for specific political reasons and from a general antipathy to peace operations. Despite recent success, peace operations remain unpopular among many Republican members of Congress. Finally, a disastrous failure might discredit peace operations and cause the United States to avoid fresh starts. A diminished level of high-end operations is illustrated in Figure 4.1.

JOINT GUARD in Bosnia-Herzegovina is scheduled to end in mid-1998, but the administration has strongly suggested that U.S. forces remain at some reduced level. If no new high-end operations were launched, U.S. effort might be limited to some follow-on in Bosnia, NORTHERN WATCH/SOUTHERN WATCH in Iraq, and two cases of traditional peacekeeping: ABLE SENTRY in northern Macedonia and the Multinational Force and Observers in the Sinai. A change of

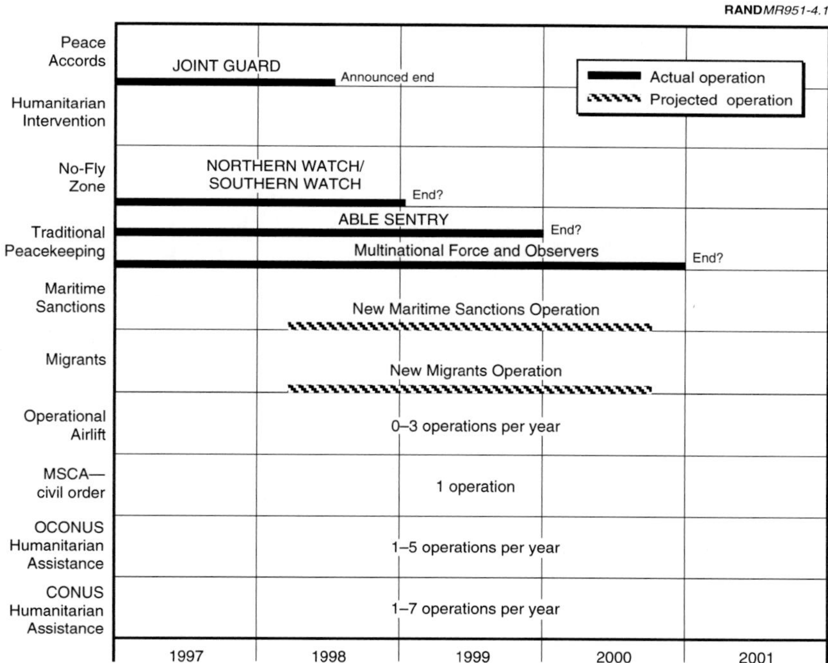

NOTE: CONUS = continental United States; MSCA = military support to civil authorities; OCONUS = outside the continental United States.

Figure 4.1—Projected Operations, 1997–2001—Diminished Level

regime in Iraq might prompt the United States to terminate the present no-fly zones.

Traditional peacekeeping tends to become interminable, but the current operations might eventually outlive their usefulness or other countries might agree to assume the burden. For example, U.N. forces might keep peace between Egypt and Israel as they did prior to the current Multilateral Force. Participation in maritime sanctions and at least some requirement to counter illegal immigration seem very likely to recur, although scope and duration are highly uncertain. Other contingency operations, especially humanitarian assistance, are likely to recur at previously observed rates.

Constant Level of Operations

Over the coming five years, the level of high-end peace operations might remain constant. There is wide agreement at senior levels in both political parties that the United States should remain a world leader and some appreciation that recurrent peacetime operations may be the price of that leadership. The humiliating failure in Somalia did not dissuade U.S. leaders from undertaking subsequent operations that were brilliantly successful, in part because lessons were learned in Somalia. UPHOLD DEMOCRACY (Haiti) and JOINT ENDEAVOR (Bosnia) are not popular, even though they have been reasonably successful to date, and they have not silenced critics of peace operations. So long as current operations proceed smoothly, the administration can ignore such criticism.

It is uncertain where conflict might erupt and prompt new peace operations, either before or after JOINT GUARD terminates. For historical reasons, the United States is unlikely to start ambitious peace operations in Southwest Asia. After the Somalia experience, U.S. leaders are extremely reluctant to become involved in central Africa, as illustrated by their very restrained response to genocide in Rwanda. On the other hand, NATO expansion might involve the United States in ethnic and national conflicts of Eastern Europe that would prompt peace operations. It is also conceivable that the United States would mount peace operations in Latin America, despite traditional fears of American "imperialism." A constant level of high-end peace operations is depicted in Figure 4.2.

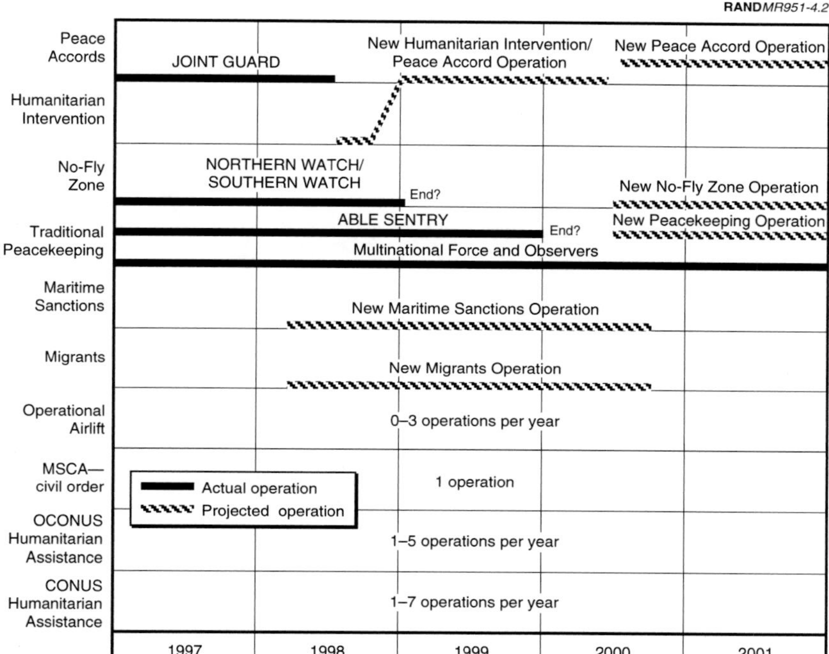

NOTE: CONUS = continental United States; MSCA = military support to civil authorities; OCONUS = outside the continental United States.

Figure 4.2—Projected Operations, 1997–2001—Constant Level

High-end peace operations after JOINT GUARD are likely to resemble those conducted in Somalia, Haiti, and Bosnia. For this analysis, we have chosen to project operations as in Somalia and Bosnia, omitting the more easily and rapidly (from a U.S. perspective) accomplished operation in Haiti. At the same time, NORTHERN WATCH/SOUTHERN WATCH might continue or be replaced by a new no-fly operation, perhaps associated with a new Bosnia-like peace operation. ABLE SENTRY and the Sinai operation might continue indefinitely or be replaced by operations of a similar type. Other contingency operations would likely continue at current rates.

Increased Level of Operations

Over the coming five years, high-end peace operations might increase, and there might also be greater demand for humanitarian assistance. There might be a major catastrophe within the United States, necessitating a far larger relief operation than previously experienced. For example, as mentioned above, Hurricane Andrew would have been far more destructive had it passed through the greater Miami area, and it would have prompted a commensurately larger relief operation. As other examples, California could experience massive movement along the notorious San Andreas Fault or the south-central United States could be devastated by movement along the Madrid Fault. Other possibilities include seismic waves (tsunamis), nuclear accidents, and even meteor strikes. In such cases, the ensuing disaster relief operation might be an order of magnitude greater than that required for Hurricane Andrew.

The level of high-end peace operations has risen over that of the previous five years, and it may not have crested. The United States might decide to conduct, almost certainly with substantial support from other countries, larger peace operations than any yet conducted. For example, Greece and Turkey might go to war over Cyprus or some issue concerning the Aegean Sea, in which case, NATO countries would exert great pressure on Greece and Turkey to conclude an early peace agreement. NATO might then undertake a very large peace operation to ensure that the agreement was implemented. An increased level of high-end peace operations is depicted in Figure 4.3.

While unlikely, this projection is still possible. Although U.S. leaders usually have considerable latitude in deciding whether to launch ambitious operations, they might also find their choices tightly constrained by circumstances, particularly since the United States seems to be the only power capable of conducting high-end peace operations successfully. U.S. leaders would have to react to a major catastrophe that occurred in the United States or its territories and would employ military forces on a larger scale than for Hurricane Andrew.

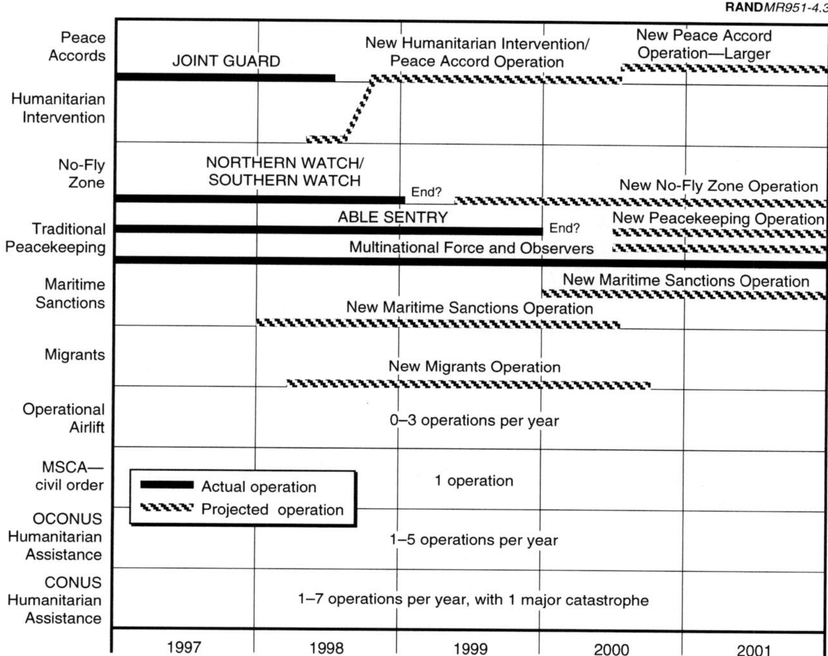

NOTE: CONUS = continental United States; MSCA = military support to civil authorities; OCONUS = outside the continental United States.

Figure 4.3—Projected Operations, 1997–2001—Increased Level

Peak Demand and Consequences

What are peak demands in the three projected futures? At a diminished level, there would be recurring demand for some specialized Army and Marine units to conduct humanitarian assistance within the United States and its territories. Small increments of airlift and sealift would be required for humanitarian assistance both within the United States and abroad. The largest demands on U.S. forces might derive from maritime operations to enforce sanctions or prevent illegal entry into this country.

At a constant level, 1–2 Army brigades[1] and one composite Air Force wing would be required for some type of high-end operation, and the Army would continue to provide two infantry battalions for traditional peacekeeping. An additional composite wing would be required to enforce a no-fly zone.

At an increased level—unlikely but not implausible—one Army division-equivalent and two Air Force composite wings would be required for a larger peace operation while another composite wing enforced a no-fly zone. One carrier battle group or surface action group would be required to enforce maritime sanctions. At the same time (indeed at any time), U.S. forces might have to respond to a major catastrophe in the United States that would require a large commitment of active forces during the initial phase and a large National Guard commitment subsequently. Demand in a single year (arbitrarily, 2001) under these three assumptions is summarized in Figure 4.4.

These projections indicate that peak demands can be met, even at an increased level. If, for example, the Army continues to maintain 10 active divisions, then its total commitment would be less than 10 percent (constant level) or 20 percent (increased level) of these forces. But there are serious difficulties in meeting these demands. First, land forces and also air forces committed to these contingency operations might not be fully or immediately available to conduct major theater warfare.[2] To preclude this risk of unavailability, the United States should have sufficient land and air forces for major theater warfare, *plus* sufficient forces to handle these contingencies.

---

[1]Commitment of combat power would be less than a full Army division, but deployment would, of course, entail more than just combat brigades. As currently constituted, Army brigades cannot deploy without appropriate divisional and corps assets to provide a wide variety of support functions. In addition, division and corps staffs would probably be tasked to provide the nucleus of joint-force headquarters.

[2]To take a recent example, the United States would not have wanted to withdraw its land forces from JOINT ENDEAVOR in order to fight elsewhere. And if it did withdraw these forces, their arrival in another theater would be delayed and their combat power might also be somewhat diminished. Air forces can redeploy with much greater ease, but they may be tied to land forces. In the same example, the United States would not wish to leave land forces in Bosnia without adequate close air support. Naval forces present a much different picture. In the same example, U.S. naval forces in the Adriatic and Mediterranean could redeploy elsewhere with no appreciable delay and without risk of failure in JOINT ENDEAVOR.

RANDMR951-4.4

	Diminished Level	Constant Level	Increased Level
Peace Accords		1 operation: 1–2 Army brigades, 1 AF wing, carrier battle group, amphibious ready group, Marine Expeditionary Unit	1 operation: 1 Army division, 2 AF wings, carrier battle group, amphibious ready group, Marine Expeditionary Brigade
Humanitarian Intervention			
No-Fly Zone		1 operation: 1 AF wing	
Traditional Peacekeeping		2 operations: one Army battalion each	3 operations: one Army battalion each
Maritime Sanctions	1 operation: carrier battle group (CVBG) or surface action group (SAG), possibly Coast Guard (CG) cutters		2 operations: 2 CVBGs/SAGs, possibly CG cutters
Migrants	1 operation: surface action group, Coast Guard cutters		
Operational Airlift	0–3 operations: usually short (days), some longer (weeks)		
MSCA—civil order	1 operation: 1–2 Army brigades, 1 National Guard (NG) division, Marine battalions		
OCONUS Humanitarian Assistance	1–5 operations: airlift, sealift, small force elements, possibly MEU		
CONUS Humanitarian Assistance	1–7 operations: battalions, squadrons, individual ships, airlift, sealift		1–7 operations: 2–5 Army brigades, 1 NG division, Marine units, airlift, sealift

NOTE: AF = Air Force; MEU = Marine Expeditionary Unit; MSCA = military support to civil authorities; NG = National Guard.

Figure 4.4—Forces for Projected Operations, 2001

Assuming that operations would continue at a constant level, these additional forces would be an important budget issue, with implications well beyond the limits of this study effort.

Second, these demands fall unevenly on the armed services and upon types of units within services. The Army and Air Force are far more seriously affected than are the Navy and Marine Corps. Navy and Marine Corps can accomplish most of their tasks within their normal deployment patterns. Reflecting their unique character, Marine forces are normally employed in the initial or concluding phases of an operation or are held in reserve; Army forces are committed for longer periods. Within the Army, some types of units are so scarce in the active force that even brigade-sized contingencies require use of National Guard and Reserve. Some types of units are so scarce in the total Army that it is difficult to rotate them or their personnel appropriately. Within the Air Force, some types of units or, more precisely, some aircraft types, are in especially high demand and short supply, making it difficult for the Air Force to maintain an acceptable level of personnel tempo for their crews. In addition, both Army[3] and Air Force[4] have learned that protracted contingencies degrade some skills through lost training opportunities.

IMPLICATIONS FOR COMMAND AND CONTROL

In the U.S. system, unified commands have relatively small forces at their disposal during normal peacetime, but they are organized and staffed to control very large forces that would be placed under their operational control during major theater warfare. To handle a smaller-scale contingency, a unified command usually generates a task force. This may be a joint task force (JTF) having more than one service component and/or a combined joint task force (CJTF) having more than one service component and more than one nation's contingent.

To conduct humanitarian intervention or coercive peace operations, a CJTF is usually needed. Moreover, a task force commander may have to interface with numerous other entities, including some that are superior and others that are autonomous or independent. These entities may include U.S. special envoys and ambassadors, Special Representatives of the Secretary-General, various agencies of the United Nations such as the United Nations High Commissioner for

---

[3]While deployed, Army units often have little opportunity to train certain skills, such as marksmanship and gunnery, so that proficiency declines and must be refreshed later. Degradation is highly differentiated by type of unit. Indeed, some support units may use their full range of skills and suffer no degradation.

[4]Instructors at the Air Force Weapons School have observed a general decline in basic maneuver skills because pilots missed training while they flew uneventful sorties over Bosnia and Iraq. See Peter Grier, "Readiness at the Edge," *Air Force Magazine*, June 1997, p. 58.

Human Rights, other international agencies such as the International Committee of the Red Cross, and nongovernmental organizations such as Co-operative for Assistance and Relief Everywhere (CARE) and Catholic Relief Services.

A unified commander usually designates an officer representing the largest service component to serve as JTF commander or sometimes CJTF commander.[5] During humanitarian intervention and coercive peace operations, the largest component is usually a land force, such as an Army division (deploying one or two brigades), a Marine Expeditionary Brigade (MEB), or a Marine Expeditionary Unit. For example, the Commander in Chief of the U.S. Pacific Command (USPACOM) designated the commanding general of III Marine Expeditionary Force (MEF), then Maj. Gen. Henry C. Stackpole III, JTF commander to accomplish SEA ANGEL in Bangladesh, because Marine forces would predominate. Stackpole's staff provided a nucleus for the JTF staff. For example, III MEF G-3 (Operations) became the JTF chief of staff, and the operations officer of 3rd Force Service Support Group became JTF J-4 (Logistics).[6] Similarly, XVIII Airborne Corps provided a nucleus for the command element of CJTF-180, and 10th Mountain Division the nucleus for JTF-190 to conduct UPHOLD DEMOCRACY in Haiti.

The staffs of unified commands may be strained by high levels of operations. For example, the U.S. European Command currently oversees operations in Bosnia-Herzegovina (JOINT GUARD), Macedonia (ABLE SENTRY), and northern Iraq (NORTHERN WATCH), in addition to recurring humanitarian operations and its own usual responsibilities.

---

[5]When one country has the lead, it may provide the nucleus of a combined headquarters, as occurred during the initial phases of operations in Somalia and Haiti. When national contingents are more nearly equal and an appropriate framework exists, the headquarters may be fully integrated, with staff positions apportioned by national contingent, as NATO operations in Bosnia are controlled.

[6]Lt. Col. (USMC) Gary W. Anderson, *Operation Sea Angel: A Retrospective on the 1991 Humanitarian Relief Operation in Bangladesh*, Annapolis, Md.: Naval War College, Strategy and Campaign Department, Report 1-92, 1992, pp. 6–7.

IMPLICATIONS FOR THE ARMY

The Army predominates in high-end operations and provides the critical capabilities. To obtain certain specialized skills and to sustain rotation of support units during protracted operations, it must access its Reserve and National Guard. As a result, individuals and units from these components have been called up and deployed more frequently than was envisioned during the Cold War. The Army has preferred volunteers whenever possible and, thus far, Reservists and Guardsmen appear to have taken these operations in stride. If, however, the operational tempo were to remain long at the current level, retention problems might ensue.

Alternative Patterns

To analyze implications for Army forces, we considered two alternative patterns: no overlap of high-end operations and simultaneous starts of such operations.

We can cite strong disincentives to simultaneous starts of high-end peace operations:

- Increase in the likelihood of strong domestic opposition
- Fuel for criticism that the United States is overreaching
- Diversion of appreciable combat forces, perhaps 10 percent of the Army's total, that would be required if some greater contingency ensued.

Therefore, as during the past five years, it seems most likely that U.S. leaders will continue to avoid simultaneous starts. Figure 4.5 projects peace operations with no overlap at the high end. The shaded area reflects a 3-year period in 6-month increments.

In this projection, the United States initiates an operation similar to RESTORE HOPE/CONTINUE HOPE (Somalia)[7] in mid-1998. Just as this operation terminates, the United States initiates an operation

---

[7]RESTORE HOPE, conducted with large U.S. forces, succeeded; CONTINUE HOPE, conducted with small U.S. forces, spiraled into failure. For this vignette, we developed force lists that would ensure success, not replicate failure.

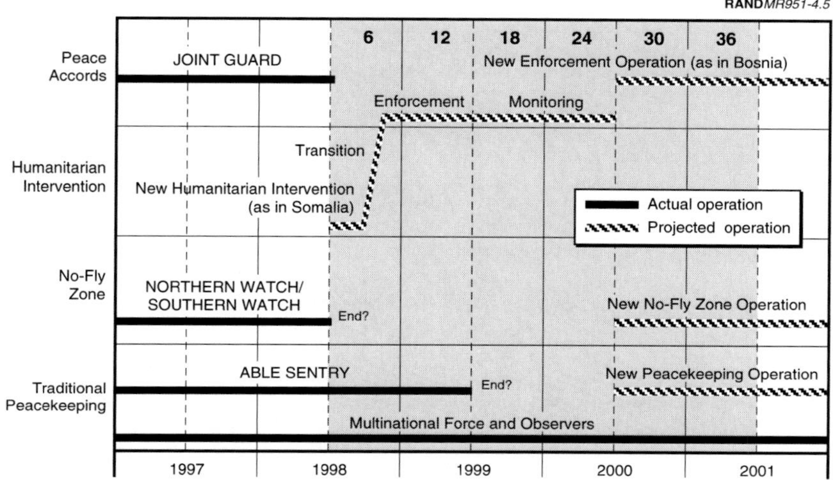

NOTE: The shaded area shows the pattern under consideration—in this case, no overlap of high-end operations from mid-1998 to mid-2001. This 3-year period is divided into 6-month increments, reflecting the usual rotation of units.

Figure 4.5—Operations, 1997–2001, No Overlap of High-End Operations

similar to JOINT ENDEAVOR/JOINT GUARD (Bosnia). At the same time, it continues to mount a no-fly operation and two traditional peacekeeping operations. The resulting projection approximates the operational tempo sustained over the past few years.

Despite obvious disincentives, the United States might initiate two high-end peace operations simultaneously or nearly simultaneously. Timing of these operations may depend on events that are partially or wholly outside U.S. control. The United States can decide whether to act, but it may not be able to choose the time. For example, during the conflict in the former Yugoslavia, the United States could not predict exactly when Croatia would launch a major offensive or whether this offensive would be successful, but the success of this offensive led directly to the Dayton Agreements that JOINT ENDEAVOR enforced. Figure 4.6 projects peace operations that overlap on the high end.

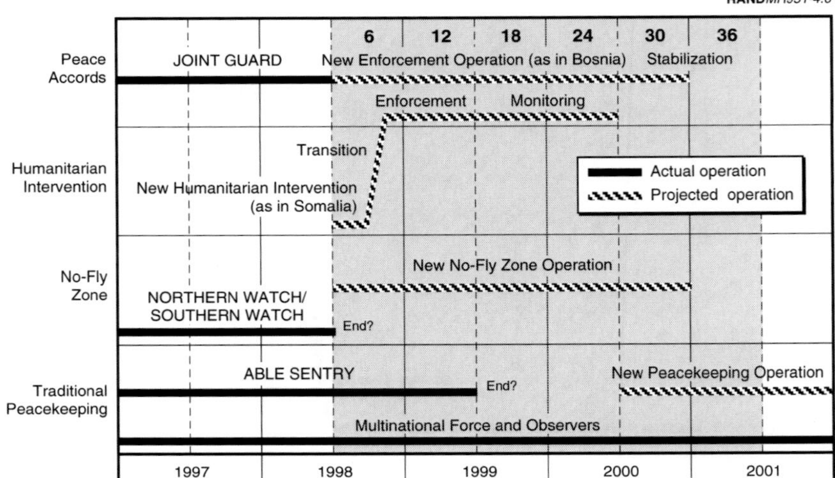

NOTE: The shaded area shows the pattern under consideration—in this case, overlap of high-end operations from mid-1998 to mid-2001. This 3-year period is divided into 6-month increments, reflecting the usual rotation of units.

Figure 4.6—Operations, 1997–2001, Simultaneous High-End Operations

In this projection, the United States initiates an operation like RESTORE HOPE/CONTINUE HOPE and an operation like JOINT ENDEAVOR/JOINT GUARD in mid-1998, just after the actual JOINT GUARD has ended. The result is a peak demand for Army forces at the beginning of a 3-year period, declining until the United States conducts only traditional peacekeeping during the last 6-month increment.

Deployed Forces

Requirements for units appear modest: Even if high-end operations were sequential, only three brigades would be required from a force that currently includes 10 active divisions. Requirements for personnel strength would also appear modest: If high-end operations were sequential, deployed personnel[8] would peak at less than 20,000; if simultaneous, at about 30,000 from an active force that currently numbers 495,000.

Although requirements appear modest measured against the active force and even smaller measured against the total Army, they cause uneven stress. The Army's structure was designed for major theater warfare, not for recurring smaller-scale contingencies. As a result, these contingencies place small demands on the Army's relatively large stock of combat forces but large demands on its relatively small stock of support units in the active force.

Requirements During Initial Phase

To analyze requirements for Army forces, we used force lists supporting the vignettes, current Army force structure, and a projection of operations at the current level.[9] If high-end operations began sequentially, the active Army could provide all required units during the initial phase. But if high-end operations began simultaneously, the active Army force structure could not provide all required units. The National Guard and Reserve would have to provide certain types of support units that are scarce in the active structure. These could include military history detachments, civil affairs units, public affairs detachments, petroleum-supply battalions, water-purification detachments, movement-control detachments, and composite group headquarters.

---

[8]"Deployed" includes only those assigned to units deployed in the area of operations (AO). It does not include personnel in support units outside the AO or personnel in units that are preparing to deploy or recovering from deployment. Personnel in support units outside the AO easily exceed those assigned to units within the AO. For example, the Army estimates that it took three soldiers stationed in the United States and Germany to support one soldier deployed in Bosnia during JOINT ENDEAVOR.

[9]For each vignette, we developed a list of Army units at battalion/separate company/detachment level, identified by Standard Requirements Code (SRC). Partial units, e.g., augmentation of a brigade staff from division staff, were expressed as percentages. Current Army force structure was drawn from the Structure and Manpower Allocation System (SAMAS), reflecting status in late 1996. In view of budget uncertainties and shifts in planning, we decided not to use projected force structure, which reflects planned activations, deactivations, etc.

Moreover, the first phase of simultaneous high-end operations would demand high percentages of certain types of units, implying that such units would be unavailable or not immediately available for other contingencies. In addition to the units noted above, the first phase could demand the entire active inventory of medium-girder bridging companies, combat aviation battalions (special operations), signal companies—tropospheric (heavy), psychological operations companies (strategic dissemination), civil affairs battalions (general-purpose), medical logistics battalions (forward), and ammunition companies (general support). It could demand about half the active inventory of combat engineer support companies, reconnaissance battalions (light division), aviation brigades (light division), attack helicopter battalions (light division), general-support helicopter battalions (light division), signal battalions (corps support), public affairs teams, surgical detachments, medical battalions (area support), preventive medicine detachments (sanitation), ammunition companies (general support—palletized load system), postal companies, light–medium truck companies, and special operations support battalions.

Requirements over Three Years

During a protracted deployment, the Army normally rotates units. The period of rotation varies according to the situation, but is typically six months. By rotating units, the Army spreads the burdens of deployment more equitably across the force and presumably minimizes retention problems that might be caused by these deployments. This practice also precludes a potential problem of dividing the Army into one portion to conduct peace operations and one portion held ready for high-intensity combat, which is seen as more demanding and more prestigious.

Only a few types of units are so scarce within the total Army that they could not be uniquely rotated, i.e., serve just one 6-month tour, during the projected 3-year period. But numerous types of units are so scarce in the active Army that they could not be uniquely rotated without calling up National Guard or Reserve units. Figure 4.7 shows selected units, Standard Requirements Code, typical personnel

RANDMR951-4.7

Unit Description	SRC	Typical Strength	Requirement as Percentage of Active Army	Requirement as Percentage of Total Army
Engineer Battalion (Light Division)	05155L	311	100	67
Attack Helicopter Battalion (Light Division)	01175L	243	100	100
General-Support Helicopter Battalion	01305A	330	140	78
Signal Battalion (Corps Support)	11445L	492	160	32
Signal Company—Tropospheric (Heavy)	11668L	60	200	40
Military Police Company (Combat Support)	19677L	183	75	5
Military Intelligence Company	34567A	51	100	100
Civil Affairs Battalion (General-Purpose)	41735L	208	260	13
Public Affairs Team	45500L	5	100	55
Medical Battalion (Area Support)	08455L	343	100	29
Medical Logistics Battalion (Forward)	08485L	226	200	80
Preventive Medicine Detachment (Sanitation)	08498L	11	113	35
Preventive Medicine Detachment (Entomology)	08499L	11	100	38
Ammunition Company (General-Support)	09633L	263	100	25
HHC Petroleum Battalion (Terminal Operations)	10416L	72	200	133
HHD Petroleum Supply Battalion	10426L	56	260	20
Water-Purification Detachment	10570L	15	240	50
Movement-Control Detachment	55580L	4	750	37
HHC Composite Group	55622L	98	240	48
Forward Support Battalion (Light)	63215L	197	80	50
Special Operations Support Battalion	63905L	467	90	90

NOTE: HHC = Headquarters and Headquarters Company; HHD = Headquarters and Headquarters Detachment; SRC = Standard Requirements Code.

Figure 4.7—Selected Units in Sequential Operations over Three Years

strengths,[10] and two entries in percentages. The first entry reflects the requirement over three years for the types of units as a percentage of the active Army, assuming 6-month rotation. Thus, "100" indicates that the active Army could satisfy the requirement if every

---

[10]Army units of identical type can be authorized different personnel strengths. These differences are usually marginal, but can be significant, especially for support units. In addition, units of identical type are routinely allocated different levels of personnel fill, i.e., are allowed to requisition different percentages of their authorized strength, according to their designated level of readiness. The strengths displayed in the accompanying figure fall in mid-range and are most frequently encountered for these types of units when fully manned.

unit of that type (not including units stationed in Korea) accomplished one 6-month rotation.

As this sample indicates, there are several different reasons for scarcity within the active force. (See Appendix C for a more complete analysis of stresses on Army units.) Some types of units, such as light divisional forces and special operations forces, are scarce because they make up a small proportion of overall Army forces, yet are particularly well suited for high-end peace operations. Some types of units, such as civil affairs units, require skills that are extremely difficult to develop and maintain within the active Army. Some types of units, especially logistics units, have been relegated to the Reserve as a matter of economy, in anticipation of mobilizing very infrequently to support major theater warfare.

Logistics Civil Augmentation Program

In 1985, the Army established the Logistics Civil Augmentation Program (LOGCAP) to facilitate contracting for logistics support and basic engineering services. In 1996, the Army Materiel Command (AMC) assumed responsibility for managing LOGCAP worldwide, a responsibility previously exercised by the Army Corps of Engineers. In recent years, sister services have initiated comparable programs: In 1995, the Navy initiated the Navy Emergency Construction Capabilities Program, essentially to provide additional engineering support following natural disasters or during contingency operations. In 1997, the Air Force initiated the Air Force Contract Augmentation Program to support forward basing during contingency operations.

Under the Army's LOGCAP concept, the winning contractor (currently, Brown and Root Services Corporation) participates in planning done by Army components of unified commands during normal peacetime. The generic planning scenario initially calls for the contractor to support 20,000 personnel in five forward locations for up to 180 days. Support includes housing, food and water, sanitation, solid-waste removal, showers, laundry, utilities, local transportation, associated construction, and maintenance of related equipment. Subsequently, the contractor may be required to support up to 50,000 personnel for up to 360 days. Since the end of the

Cold War, the Army has used LOGCAP in the operations listed in Figure 4.8.

The Army considers LOGCAP a last resort for when other options are unavailable or unacceptable—for example, using military assets, accepting host-nation support, and contracting with local suppliers. In large operations, such as JOINT ENDEAVOR, LOGCAP can

- maximize combat troops under ceilings
- minimize Reserve and National Guard call-up
- employ less-expensive local labor
- provide higher quality of life.

RAND*MR951-4.8*

Operation	Support Provided Under LOGCAP	M$
RESTORE HOPE (Somalia, 1992–1993)	Base-camp construction, food and water, sanitation, solid-waste removal, showers, laundry, utilities, bulk fuel handling, local transportation, linguist support.	110
SUPPORT HOPE (Rwanda, 1994)	Water production, storage, and distribution.	6
UPHOLD DEMOCRACY (Haiti, 1994–1995)	Base-camp construction, food and water, laundry, bulk fuel handling, local transportation, linguist support, airport and seaport operations.	141
VIGILANT WARRIOR (Kuwait and Saudi Arabia, 1994)	Food and water, laundry, local transportation, seaport operations (container handling).	5
DENY FLIGHT (Italy, 1995–1996)	Base-camp construction.	6
JOINT ENDEAVOR (Bosnia, 1995–present)	Base-camp construction, food and water, sanitation, solid-waste removal, showers, laundry, bulk fuel handling, local transportation, mail delivery, railhead operations, seaport operations.	462

SOURCE: David R. Gallay and Charles L. Horne III, *LOGCAP Support in Operation Joint Endeavor: A Review and Analysis*, McLean, Va.: Logistics Management Institute, LG612LN1, 1996, pp. 11–15; and U.S. General Accounting Office, *Contingency Operations: Opportunities to Improve the Logistics Civil Augmentation Program,* Washington, D.C., GAO/NSIAD-97-63, February 1997, p. 7. Costs are contract value in millions of dollars according to Gallay and Horne.

NOTE: The cost for JOINT ENDEAVOR, an ongoing operation, is estimated.

Figure 4.8—Use of LOGCAP

As a matter of policy, the United States often limits operations by setting a troop ceiling. For example, during JOINT ENDEAVOR, the troop ceiling for Bosnia-Herzegovina was set at 20,000 personnel. By using LOGCAP, the Army avoided sending thousands of troops in support units and sent combat troops (plus other needed specialties) instead. LOGCAP allows the Army to minimize call-up of Reservists and National Guardsmen who might otherwise be needed in support units. Viewed from a related perspective, LOGCAP allows the Army to make better use of whatever call-up is authorized by concentrating on the most-needed specialties, such as civil affairs. Moreover, local labor provided under LOGCAP works at wage rates far lower than active-duty pay for Reservists or National Guardsmen. Logistics Management Institute (LMI) calculated that the Army would have spent over $200 million in additional active-duty pay had it used military units rather than LOGCAP during JOINT ENDEAVOR.[11] Finally, LOGCAP can offer a higher quality of life than the Army would normally provide for itself under field conditions. When the Army goes into the field, it usually dispenses with amenities, such as some food services, laundry, and solid-waste disposal, that are provided under civilian contract in-garrison. This lower quality of life is unobjectionable for short periods—for example, during training exercises—but becomes questionable when endured for six months at a time.

Using LOGCAP has some drawbacks and limitations:

- During an initial phase of deployment, military forces may have to provide essential lift and engineering support to get LOGCAP started, especially if infrastructure is damaged or absent.
- LOGCAP activities may require protection by military forces.
- LOGCAP personnel cannot be expected to take up weapons in a crisis, as personnel in military support units would.

---

[11]To analyze relative cost, LMI developed a comparable military support package that included one active unit of each type, with the remaining units selected from the Reserve and National Guard. Disregarding pay for personnel in the active Army (who would be on active duty anyway), LMI calculated that personnel in this support package would have cost $318 million in active-duty pay, assuming 6-month rotation with two months to prepare and one month to stand down. In contrast, labor costs under LOGCAP were only $100 million, employing a workforce that was over 80 percent foreign nationals. Gallay and Horne, 1996, pp. 23–25.

- The Army has experienced recurrent problems in managing LOGCAP. These problems can be traced to planning uncertainties and inexperience in negotiating costs. AMC expects to solve these problems by providing training in LOGCAP and establishing LOGCAP support teams in the Army components of unified commands.

IMPLICATIONS FOR THE AIR FORCE

The Air Force, supplemented by commercial carriers, has primary responsibility for strategic and in-theater airlift. Since the end of the Cold War, the Air Force has sustained a high level of effort to deliver humanitarian aid, to deploy forces, and to sustain forces in-theater. These operations have often required that overseas bases be augmented and austere basing be developed. The Air Force has met this challenge largely from its own resources, with some assistance from sister services.

The Air Force predominates in enforcement of no-fly zones and strike missions, although the Navy also contributes. Whether conducted in isolation (e.g., Iraq) or in support of land operations (e.g., Bosnia), no-fly zones have lasted for years, making large cumulative demands on the Air Force. Protracted operations have placed especially great stress on the Air Force's small fleets of specialized aircraft that perform air-defense suppression, electronic warfare, reconnaissance, and recovery. Despite routine use of Reserve and Air National Guard assets, the Air Force has experienced difficulty keeping temporary duty (TDY) to 120 days per year. High rates of TDY are one cause of falling retention rates for pilots. Unless the Air Force can reverse this current trend, it will soon begin to experience spot shortages of trained pilots.

Basing Challenges

The Air Force has flown most combat missions from well-established air bases. For example, missions to support JOINT ENDEAVOR (Bosnia) were flown primarily from air bases in Italy: Aviano (USAF: OA-10, F-16, EC-130; USMC: F/A-18), Brindisi (USAF: AC-130, MH-53, MC-130; USMC: EA-6B), and Pisa (USAF: KC-135). (The Navy has periodically conducted carrier operations in the Adriatic Sea.

NATO air forces have operated from Aviano, Brindisi, Cevia, Ghedi, Gioia del Colle, Istrana, Piacenza, Pisa, Pratica di Mare, Rimini, Vicenza, and Villafranca.) Air-denial operations over northern Iraq were flown from Incirlik Air Base near Adana, Turkey; those over southern Iraq were flown largely from Dhahran in Saudi Arabia until an attack on Air Force personnel[12] prompted the decision to concentrate operations at a more defensible facility: Prince Sultan Air Base near Al Khari, southeast of Riyadh. Until it was expanded into an air base by Air Force engineers, Prince Sultan was merely an airstrip, built during DESERT SHIELD and unused since the Gulf War.

The Air Force has flown many airlift missions into austere bases, both to deliver humanitarian supplies and to deploy U.S. forces. To support airlift, Air Mobility Command (AMC) deploys Tanker Airlift Control Elements (TALCEs) to CONUS and overseas bases as required. TALCEs are provisional organizations with no definite composition. They are task-organized according to mission requirements and conditions at the proposed bases. At one extreme, a TALCE might be limited to control elements, as the name suggests. At another extreme, a TALCE might include elements to provide aerial port services, aeromedical evacuation, civil engineering, communications, equipment maintenance, intelligence, ground transportation, logistics support, postal service, refueling, and weather forecast.

During UPHOLD DEMOCRACY, AMC deployed TALCEs to Borinquen, Puerto Rico; Cap-Haïtien, Haiti; Cecil Field Naval Air Station, Florida; Homestead Air Base, Florida (since closed); MacDill Air Base, Florida; Opa-Locka Airport, Florida; Port-au-Prince, Haiti; and Roosevelt Roads, Puerto Rico. During SUPPORT HOPE, AMC deployed TALCEs to Addis Ababa, Ethiopia; Entebbe, Uganda; Mombasa, Kenya; Goma, Zaire; Harare, Zimbabwe; Kigali, Rwanda; and Nairobi, Kenya. During JOINT ENDEAVOR, AMC established TALCEs at Aviano, Italy; Belgrade, Serbia; Brindisi, Italy; Budapest, Hungary; Gulfport, Mississippi; Pisa, Italy; Ramstein Air Base,

---

[12]On June 25, 1996, unknown perpetrators detonated a large truck-bomb near Khobar Towers in Dhahran, killing 19 Air Force personnel and wounding hundreds of others.

Germany; Rhein-Main Air Base, Germany; Taszar, Hungary; Tuzla, Bosnia; and Zagreb, Croatia.[13]

TALCEs may include an engineer squadron (RED HORSE) if heavy engineering is required. RED HORSE squadrons provide essential engineering support, including heavy earthmoving (clearance, grading, revetments), rapid runway repair, road construction, erection of impermanent facilities, power generation, well drilling, and explosives demolition. They are configured and packaged for early deployment, either as complete units or in teams. The smallest RED HORSE team (RH-1) consists of 16 personnel without heavy equipment. Within 12 hours of notification, this team can deploy to a remote site, where it will conduct an airfield survey and prepare a beddown plan. Two active RED HORSE squadrons are located in CONUS: one at Nellis Air Force Base (AFB), Nevada, and one at Tyndall AFB/Hurlburt Field, Florida. Two active RED HORSE flights are located abroad: one at Leghorn, Italy, and one at Osan Air Base, Republic of Korea. Four RED HORSE units are in the Air National Guard and two are in the Air Force Reserve. In addition, all civil engineer squadrons are organized into Prime Base Engineer Emergency Force (PRIME BEEF) lead and follow-on teams. Depending on the situation, PRIME BEEF can deploy in small teams or larger increments of up to 200 personnel. Engineer support can also be provided by sister services: Army engineer units, Naval Mobile Construction Battalions (NMCB), and Marine Corps engineer units.

Modularity

The limited scope of most airlift, no-fly zone, and strike operations means that air forces have usually deployed in relatively small increments. For combat units, these increments have been squadrons and individual air crews, usually organized into a provisional wing. For support units, these increments have been modules and individuals, organized under a provisional wing for combat operations or under a TALCE to support airlift. For example, during SUPPORT HOPE, the 621st TALCE deployed to Entebbe, Uganda; it included elements drawn from two aerial port squadrons, two air-movement-

---

[13]U.S. Air Force, Headquarters, Air Mobility Command, *1997 Air Mobility Master Plan (1997-AMMP)*, Scott AFB, Ill., 1996, pp. 2-15 through 2-17.

control squadrons, an air logistics squadron, a civil engineering squadron, an equipment-maintenance squadron, a mission-support squadron, and an operations-support squadron.

To facilitate these incremental deployments, the Air Force employs modules designated by unit-type codes (UTC). Each UTC designates an element that can be taken from a parent unit. In this way, Air Force units are permanently tasked to prepare for deployment by UTC rather than by unit. For example, aerial port squadrons have UTCs designating elements that can support unit moves, provide cargo services at 75 tons/day for three aircraft at maximum on ground (MOG 3), and handle wide-body aircraft load. Civil engineer squadrons have UTCs designating PRIME BEEF lead teams, PRIME BEEF follow-on teams, and explosive ordnance threat-augmentation teams. Security police squadrons have UTCs designating security police flights, 81mm mortar teams, .50-caliber machine-gun teams, Mark 19 grenade teams, fire direction center teams, etc. Parent units can give up some UTCs while remaining fully capable, although at a lower level of effort; other UTCs may imply losing ability to perform some functions at all.

Air Expeditionary Force

The Air Force developed an Air Expeditionary Force (AEF) to deter Iraqi aggression and conduct air-denial operations in southern Iraq (SOUTHERN WATCH).[14] An AEF is built around a designated wing that provides the command element. It typically comprises 30–40 aircraft drawn from 3–4 wings, including 12 F-15C/Ds and/or F-16s for air superiority, 12 F-15Es and/or F-16C/Ds equipped with Low-Altitude Navigation and Targeting Infrared for Night (LANTIRN) for ground attack, 6 F-16s with High-Speed Antiradiation Missiles (HARMs), and other support aircraft, such as KC-135s for aerial refueling. In addition, 6 B-1s and/or B-52s are committed to an AEF but remain stationed in CONUS.

---

[14]In October 1994, U.S. Central Command (USCENTCOM) rapidly deployed forces to the Persian Gulf (VIGILANT WARRIOR) to counter a threatening Iraqi concentration. Deployment of land-based air forces proved more difficult than expected, prompting the Air Force to develop the AEF concept. Since then, USCENTCOM has deployed AEFs four times to conduct SOUTHERN WATCH: twice to Qatar, and once each to Bahrain and Jordan.

In its current form, the AEF concept requires bases where equipment and supplies may be prepositioned during peacetime. When the deployment order is given, an advance team deploys to open the prepositioned sets and prepare for flight operations. With this advance team are combat pilots who plan initial sorties while awaiting the arrival of combat aircraft that self-deploy using aerial refueling. When these aircraft arrive, their pilots are exhausted from hours spent in cramped cockpits; but the early-arriving pilots are rested and ready to initiate combat sorties as soon as the planes are checked, refueled, and armed. The announced goal, so far not achieved in practice, is to deploy an AEF and start generating combat sorties within 48 hours of receiving an order to execute.

The Air Force is continuing to experiment with AEFs, using the 366th Wing located at Mountain Home AFB, Idaho. The 366th is a composite wing equipped with F-15C/D/E, F-16C/D, KC-135R, and B-1B aircraft—in other words, a permanent AEF. The Air Force had intended to create more composite wings like the 366th, but operations and maintenance costs for unlike aircraft proved too expensive. The AEF concept keeps aircraft in pure wings that realize economies of scale while preparing to form composite wings when they are needed. Up to 1997, AEFs had been optimized for combat, i.e., air superiority and ground attack, but the Air Force may apply the concept to other missions such as airlift to provide humanitarian aid.

Uneven Stress on Aircraft and Crews

As with Army forces, deployed air assets appear modest compared with the available pool. Since 1992, from 1.5 to 2.2 fighter wing equivalents (FWEs) have been continuously required to enforce no-fly zones and to conduct strikes over Bosnia and Iraq, as compared with 13 FWEs in the active Air Force[15] and another 8 FWEs in the Reserve and Air National Guard. Even an increased level of operations, projected to require 3 composite wings, could be sustained from the active Air Force, although tours would exceed the 120-day ceiling set as a goal by the Air Force Chief of Staff.

---

[15]During early 1997, the context of the Quadrennial Defense Review (QDR), the Air Force accepted reduction of active strength to 12 FWEs in order to fund modernization, especially the F-22 program.

But these apparently modest requirements mask very uneven stresses on the force. To discern implications for Air Force assets, we used data at the level of aircraft type, current Air Force structure, and a projection of operations at current level. Data on squadron deployments is unhelpful, because aircraft types are deployed much below squadron strength. Even if deployment were by full squadrons, such data would still require expansion to account for the variation in primary aircraft authorized (PAA), e.g., F-15/F-16 squadrons may have 18 or 24 aircraft; MH-53J squadrons may have 5 or 22 aircraft.

Air-denial operations demand small numbers of aircraft specialized in air-to-air combat and strike relative to large numbers in the force, but large numbers of certain supporting aircraft relative to the small numbers in the force. The reason is easy to discern: These operations entail little combat, yet they extend over wide areas of operations that generate theater-like demands for specialized functions, such as air-defense suppression, electronic combat, reconnaissance, and recovery. They also make fairly heavy demands on air refueling, both for combat aircraft and for special operations aircraft tasked with recovery of downed air crews.

Peak demand during DELIBERATE FORCE, PROVIDE COMFORT (Northern Iraq), and SOUTHERN WATCH (Southern Iraq) involved fairly high percentages of the active force and stressed the force unevenly. At peak, the Air Force deployed half of the EF-111s,[16] over one-third of the EC-130s and RC-135s, about one-quarter of the A-10s and E-3s, and one-fifth of the HH-60s from the active force. It also deployed 15–17 percent of the F-16 fighters and the KC-10/KC-135 refueling aircraft. (See Figure 4.9.)

If operations remain at the current level, implying that the Air Force will have to support one peace-accord operation and to conduct one no-fly operation simultaneously, stress will continue to be very un-

---

[16]Since 1981, the Air Force has employed EF-111A aircraft to suppress air defense by electronic means, such as jamming acquisition radars. EF-111A aircraft recently accomplished this mission over Bosnia and Iraq, but they are being phased out pursuant to an agreement between the Air Force and Navy for both services to rely on the Navy's EA-6B Prowler.

RANDMR951-4.9

Aircraft Type/ Name	Function	Component			Operation			Total Number	Percentages	
		AC	ANG	AFR	DELIBERATE FORCE	PROVIDE COMFORT	SOUTHERN WATCH		AC	AF
EF-111 Raven	Defense suppression	26			6	3	4	13	50	50
A/OA-10 Thunderbolt	Close attack/FAC	118	104	44	12		24	36	31	14
F-15 Eagle	Fighter	549	119		8	6	30	44	8	7
F-16 Falcon	Fighter	705	654	32	46	18	41	105	15	8
F-117 Nighthawk	Stealth fighter	47					8	8	17	17
M/HH-60G Pave Hawk	Search and rescue	35	18	25		2	5	7	20	9
E-3 Sentry	Warning and control	29				3	4	7	24	24
EC-130 (special missions)	Special duty	16	6		6			6	38	27
RC-135 Rivet Joint	Reconnais-sance	13			2		3	5	38	38
U-2	Reconnais-sance	28			3		2	5	18	18
AC-130 Spectre	Gunship	19			4			4	21	21
MC-130 Combat Talon	SO transport	46	4	12	2			2	4	3
MH-53 Pave Low	SO helicopter	36			4			4	11	11
HC-130 Combat Shadow	SO tanker	8	4	14		2	2	4	50	15
KC-10 Extender	Air refueler	54					9	9	17	17
KC-135 Stratotanker	Air refueler	228	224	62	12	5	17	34	15	7

NOTE: AC = active component; AF = Air Force (total); AFR = Air Force Reserve; ANG = Air National Guard; FAC = forward air controller; SO = special operations.

Figure 4.9—Peak Demand for Air Forces

evenly distributed. These operations would probably require only 5–10 percent of the Air Force's fighter and ground attack aircraft, but from one-quarter to one-third of some aircraft specialized in electronic combat, reconnaissance, and recovery. Nor could the Air Force Reserve Command and Air National Guard do much to relieve stress on these specialized aircraft. (See Figure 4.10.)

RANDMR951-4.10

Aircraft Type/ Name	Function	Component			Operation		Total Number	Percentages	
		AC	ANG	AFR	New Peace Accord	New No-Fly Zone		AC	AF
EF-111 Raven	Defense suppression	26			6	6	12	46	46
A/OA-10 Thunderbolt	Close attack/FAC	196	101	45	12		12	6	4
F-15 Eagle	Fighter	549	116		6	24	30	5	5
F-16 Falcon	Fighter	705	631	114	24	24	48	7	3
F-117 Nighthawk	Stealth fighter	47				6	6	13	13
M/HH-60G Pave Hawk	Search and rescue	35	18	25	2	6	8	23	10
E-3 Sentry	Warning and control	29				6	6	21	21
EC-130 (special missions)	Special duty	16	6		3	3	6	38	27
RC-135 Rivet Joint	Reconnais-sance	15			2	2	4	27	27
U-2	Reconnais-sance	28			2	2	4	14	14
AC-130 Spectre	Gunship	19			4		4	21	21
MC-130 Combat Talon	SO transport	46	4	12	4		4	9	6
MH-53 Pave Low	SO helicopter	36			4	4	8	22	22
HC-130 Combat Shadow	SO tanker	8	4	6	2	2	4	50	22
KC-10 Extender	Air refueler	54		62	3	3	6	11	5
KC-135 Stratotanker	Air refueler	228	224	63	12	18	30	13	6

NOTE: AC = active component; AF = Air Force (total); AFR = Air Force Reserve; ANG = Air National Guard; FAC = forward air controller; SO = special operations.

Figure 4.10—Current-Level Demand for Air Forces

Looming Retention Problem

The Air Force is losing so many pilots that it may soon fall short of requirements. Over the past three years, the rate at which pilots were willing to take the pilot bonus ("take rate") and thus continue duty beyond their initial commitment of eight years' active service has declined and is now running well below 50 percent. A primary cause is high demand from airlines, including the increasingly attractive regional carriers. But, in response to a recent survey, high operational tempo was the most frequently cited reason for declining the bonus.[17]

To address the looming retention problem, the Air Force has taken several courses of action:

- As a matter of policy, units returning from contingencies are to stand down, i.e., to curtail normal activities, for a week so that members can devote time to their families.
- The Air Force has reduced the number of staff positions that require pilots, so that more will be available to fly.
- The Air Force has also sought to reduce the level of effort in contingency operations—for example, the frequency of combat air patrols flown routinely to enforce no-fly zones.
- The Chief of Staff of the Air Force has set a goal to keep TDY, including deployment for training and exercises, from exceeding 120 days per year.

The last goal is especially problematic. At the current level of operations, the Air Force has difficulty meeting the 120-day goal, even for fighter aircraft, the largest category.[18] Depending on location, active-duty fighters currently need 50–60 days for TDY other than contingencies, such as routine training and joint exercises, which leaves 60–70 days for contingencies. Fighters in the reserve components need about 35 days for training, which leaves 15 days for contingencies. Therefore, the Air Force requires about 4 U.S.–based aircraft or 5 European-based aircraft or 23 reserve-component aircraft to support each deployed aircraft at no more than 120 days' TDY, assuming that aircraft deploy with the same crew ratio as at home station. By these calculations, current Air Force structure can support contingencies involving just 2.2 FWEs on a continuing basis without exceeding the 120-day goal. Of this total, the reserve components contribute slightly less than a squadron-equivalent.

---

[17]Thirty percent cited high operational tempo; 17 percent cited other quality-of-life issues; and 15 percent cited desire to work for an airline as their reason for declining. See Suzanne Chapman, "Keeping Pilots in the Cockpits," *Air Force Magazine*, July 1997, p. 69.

[18]This analysis of temporary-duty rates is drawn from unpublished work by RAND colleague David Thaler within Project AIR FORCE.

IMPLICATIONS FOR THE NAVY AND MARINE CORPS

The Navy predominates in preventing illegal migration by sea and in enforcing maritime sanctions. Through carrier operations, it also makes significant contributions to enforcing no-fly zones and conducting strikes. With the exception of the maritime sanctions against Iraq that preceded the Gulf War, Navy forces have conducted these operations within their normal deployment patterns. The Marine Corps has usually conducted operations with an embarked Marine Expeditionary Unit (MEU). Exceptions have been SEA ANGEL, conducted by a Marine Expeditionary Brigade (MEB) in transit from the Persian Gulf War, and RESTORE HOPE. Neither service has experienced stresses that would have implications for force structure.

Marines are usually employed in a Marine air-ground task force (MAGTF). The smallest customary MAGTF is the MEU, built around a reinforced Marine infantry battalion. An MEU is usually transported by a three-ship ARG, typically including one ship in the Tarawa class (LHA) or Wasp class (LHD), one ship in the Harper's Ferry class (LSD) or Whidbey Island class (LSD), and one ship in the Shreveport class (LPD) or San Antonio class (LPD). Wasp- and Tarawa-class ships displace approximately 40,000 tons. They have flight decks and below-deck storage for Harriers and helicopters, plus well decks to launch landing craft. Harper's Ferry– and Whidbey Island–class ships displace about 18,000 tons. They have helicopter-landing pads, well decks, cargo spaces, and on- and off-loading equipment including heavy cranes. Shreveport-class ships, due to be replaced by the larger San Antonio class, displace approximately 17,000 tons. They have a helicopter-landing pad, a smaller well deck, and on- and off-loading equipment. The more-capable San Antonio–class ships will operate CV-22 aircraft as well as helicopters.

The Marine Corps has six MEUs: 11th, 13th, and 15th MEUs based at Camp Pendleton, California; 22nd, 24th, and 26th MEUs based at Camp Lejeune, South Carolina. The command elements are permanent organizations. Units are assigned to these commands on a rotational basis, which includes six months of sea duty (a "pump," in Marine jargon). At any given time, two MEUs are deployed operationally; two are in pre-deployment training; and two are in some administrative status, including transit to an operational area. Marine Expeditionary Units are designated as Special Operations

Capable after training that includes noncombatant evacuation, recovery of personnel and aircraft, hostage rescue, clandestine reconnaissance, special demolitions, amphibious raids, and combat in urban terrain.

An MEU includes a ground combat element, an aviation combat element, and a combat service support element. The ground combat element is a reinforced Marine infantry battalion, usually three infantry companies, a reconnaissance platoon, a weapons company, an artillery battery (105mm howitzers), a tank platoon (M-60 main battle tanks), a light armored platoon, a combat engineer platoon, and an assault amphibian platoon. It also includes a fire-control party drawn from an air and naval gunfire liaison company (ANGLICO) and intelligence assets. The aviation combat element is built around a medium-helicopter squadron. It usually includes AV-8B Harriers, AH-1W Sea Cobra attack helicopters, and CH-53 Sea Stallion, CH-46 Sea Knight, and UH-1 transport helicopters. The combat service support element provides communications, beach-landing support, engineer support, maintenance, medical services, transportation, and resupply.

There is a natural division of labor between Marine forces and Army forces. Marine forces are optimized to force entry on a littoral, as during RESTORE HOPE in the Mogadishu area and UPHOLD DEMOCRACY at Cap-Haïtien. In both cases, the parties offered no resistance, largely because they realized it would have been futile. During UNITED SHIELD, Marine forces re-entered Somalia to extract a U.N. force that could not secure its own departure. Army forces are better suited to conduct protracted operations inland, including associated sustainment. Thus, Army forces assumed responsibility for the protracted phase of RESTORE HOPE and CONTINUE HOPE in Somalia, and for protracted operations during UPHOLD DEMOCRACY and RESTORE DEMOCRACY in Haiti.

As an interesting exception to this division of labor, 24th MEU (SOC) operated far inland during PROVIDE COMFORT I. On April 13, 1991, 24th MEU (SOC) landed at Iskenderun, Turkey, and began to establish a forward base at Silopi, a village 450 miles to the east, in the Taurus Mountains. On April 25, the Marines occupied the town of Zakho in northern Iraq (Kurdistan). They left northern Iraq on July 15 and shipped out of Iskenderun on July 19, some three months after they had landed. During this operation, 24th MEU (SOC) controlled and, to some degree, supported a 3,600-man international brigade, straining the small Marine force to its limit.[19] 24th MEU (SOC) was reinforced during the operation by additional ANGLICO fire-control teams to support other U.S. and allied forces and an Unmanned Aerial Vehicle Detachment to give video coverage of roads and Iraqi positions.

---

[19]Ronald J. Brown, *Humanitarian Operations in Northern Iraq, 1991, with Marines in Operation Provide Comfort*, Washington, D.C.: Headquarters, U.S. Marine Corps, History and Museum Division, 1995, p. 69.

Chapter Five

RECOMMENDATIONS

In the final phase of the project, we developed options—including modifications to force structure and procedural changes—to improve the conduct of humanitarian intervention and coercive peace operations without detracting from the nation's ability to prevail in major theater warfare.

In recent years, U.S. forces have conducted these operations with impressive success, especially at the tactical level—a record demonstrating that they already have the required capabilities and, therefore, that large changes are not required.

The essential U.S. contribution to these operations is power projection, the ability to rapidly deploy and sustain forces prepared for high-intensity combat. There is no reason for the United States to develop less-capable constabulary forces that other countries could provide as well. Indeed, U.S. force requirements for these operations and for outright interventions, such as URGENT FURY or JUST CAUSE, are almost identical. Accordingly, most of the options proposed in this study would contribute to power projection broadly defined, not just to better conduct of humanitarian intervention and coercive peace operations.

The frequency and size of these operations are highly uncertain and might decline to a Cold War level. Faced with such uncertainty, a prudent planner would like to select those options that are less sensitive to the level of operations. Most options considered in this report would increase warfighting capability and therefore be advisable even if the level of these operations declined. Others would occasion little regret if the level declined. Only a few might decrease warfighting capability and therefore occasion regret if the level declined.

In this chapter, we discuss the options we developed, one option per section. Each section evaluates the option in relation to the level of future operations, resource commitment, and power projection.

REFINE COMMAND ELEMENTS FOR CJTF

The Joint Staff, unified commands, and services further refine those command elements that are required for a combined joint task force (CJTF). CJTF command elements are organized according to a standard pattern, and their officers are predesignated on a contingency basis.

This option is insensitive to assumptions about the level of future operations. It entails only modest commitment of resources, largely the man-hours associated with planning and exercising command elements. Moreover, it would improve power projection for other purposes, whether conducted unilaterally or in concert with other countries.

Predesignations would reflect normal peacetime staffing of command elements that might form the nuclei of CJTF—e.g., Army corps/divisions, Air Force numbered air forces, Navy fleets, Marine Expeditionary Brigades. CJTF command elements would "stand up," i.e., assemble and become active, then exercise frequently enough to ensure that the predesignated officers are proficient despite the rotational cycle of normal assignments. During exercises, staffs might assemble in one location, such as a wargaming facility, or they might network from several locations; predesignated officers would also become directly acquainted with forces of sister services. Exercises would include play with foreign forces and with those civilian agencies that are frequently involved in humanitarian intervention and peace operations—e.g., U.S. Department of State, U.S. Department of Justice, U.S. Department of Agriculture, U.S. Agency for International Development, United Nations High Commission for Refugees, and the International Committee of the Red Cross.

PERFORM SEARCH AND RESCUE USING A VARIETY OF FORCES

Unified commands employ a variety of specialized forces to perform search and rescue in denied areas, including those forces subordinate to the U.S. Special Operations Command (USSOCOM) and other forces within the services.

This option would help to even the stress on specialized forces if the level of operations remains at least constant. Search and rescue during air-denial and strike operations puts heavy demands on specialized forces within the active components of the services, especially certain aircraft (M/HH-60, MC-130, MH-53, HC-130) and their crews. The requirement is determined by level of threat and geographic extent of air operations. In some situations, such as rescue of pilots and crews enforcing no-fly zones from Iraq under the rule of Saddam Hussein, the best-trained and best-equipped forces may be required. In other situations, requirements may be less stringent.

To even stress, unified commands should spread the burden of search and rescue over all specialized forces that are appropriate in the prevailing situation. These forces may include Army Special Forces, Rangers supported by Army aviation, and MEU (SOC).

EXPAND USE OF CIVILIAN CONTRACTORS

The services expand use of civilian contractors to support contingency operations ranging from disaster relief to coercive peace operations.

This option is insensitive to the level of operations; it would merely be exercised less frequently if the level declines. The Army's LOGCAP and comparable programs in sister services do not obviate requirements for support units that perform similar functions. The services still require military support units that will continue to perform their duties under conditions intolerable to civilian organizations. But peace operations usually imply more tolerable conditions, which allow use of civilian contractors, even during initial phases. Use of contractors lessens call on inactive components and can save money by employing local labor at a rate lower than U.S. active-duty pay. It can also contribute to reviving economic life, which furthers a peace process.

DEFINE STRUCTURE OF ARMY CONTINGENCY BRIGADES

The Army defines the structure of Army contingency brigades that would be activated when the need arose.

This option would do no harm if the level of operations declines and would be highly beneficial if the level remains at least constant. At little cost, it makes Army forces a more effective and better-understood instrument of national power in a broad range of contingencies, including unilateral interventions. The Army could gain the following advantages:

- Army forces would deploy quickly on short notice.
- Army forces would operate more efficiently, especially during the critical first phase of an operation.
- National Command Authority and JTF commanders would better understand and appreciate Army capabilities.

Under current organization, the brigade is an operational level of command that requires support (administrative, logistics, medical, etc.) from division and corps. If a brigade has to operate independently, slices of higher-echelon assets must be provided. Providing these on an *ad hoc* basis causes some turmoil and initial uncertainty. (An exception would be the ready brigade of the 82nd Airborne Division, which routinely prepares for independent operations.) Of course, exigencies of different theaters and missions may require variations in the slices, but recent practice indicates that the broad requirements are well understood and fairly stable. Organizing contingency brigades with integral support would promote efficiency, especially during deployment and the initial phase of operations.

The Army task-organizes forces at battalion- and brigade-levels in a highly flexible way. For smaller-scale contingencies, it usually generates organizations uniquely tailored to the expected operations. Although flexibility is desirable, such a high degree of flexibility carries penalties. Developing the task organization takes time, and

Army personnel must become acquainted with the resulting command arrangements and capabilities. The learning experience is more difficult for joint staffs that are confronted by a unique organization for the first time and have no experience in planning and controlling its actions.

Force requirements for the initial phase of a smaller-scale contingency are quite predictable. During the initial phase, U.S. forces must show overwhelming combat power to deter opposition, implying brigade-sized task forces prepared for intense combat. These brigades would be equally well suited to conduct peace enforcement, e.g., JOINT ENDEAVOR, or unilateral interventions, e.g., URGENT FURY. They would be no more closely associated with peace operations than are Marine MEU and MEB. During subsequent phases of an operation, Army forces could be adjusted to accomplish such collateral missions as reconstruction of infrastructure or support for electoral activities.

In this option, the Army would define the structure of contingency brigades by types of units and subunit increments, establish planning factors for deployment of brigades, and develop doctrine for their employment. The Army might also designate specific units, much as units are identified for planning purposes in current war plans. If specific units were identified, the Army would have to guard against the danger of dual command, i.e., command through both Army channels and JTF channels during normal peacetime. The Army would also organize and exercise expanded brigade staffs that could form the nucleus of JTF headquarters.

An Army contingency brigade might include the following elements (see Appendix D for a notional force list):

- Command element: heavy brigade headquarters augmented from divisional and corps staff, intelligence assets
- Maneuver element: armor battalion, mechanized infantry battalion, light/motorized infantry battalion, field artillery battalion, air-defense company, special forces company, military police company
- Aviation element: medium-helicopter battalion, attack-helicopter battalion, elements from an aviation support battalion

- Support element: forward support battalion, medical battalion, ammunition company, aviation maintenance company, 2 signal companies, 2 truck companies, petroleum supply company, ordnance team, civil affairs team.

USE NATIONAL GUARD AND RESERVE FOR NONCOERCIVE PEACEKEEPING

The Army uses National Guard and Reserve forces for noncoercive peacekeeping as a matter of policy.

This option is almost insensitive to assumptions about the level of future operations. At a diminished level, it tends to become irrelevant; at constant or higher levels, it becomes worthwhile. In traditional peacekeeping, a military force monitors compliance with an agreement, reports violations, and attempts to resolve violations. This military force is not expected to enter combat, except in self-defense if attacked. This mission is well within the capability of National Guard and Reserve units and individual members.

Employment of National Guard and Reserve would offer several advantages:

- Less diversion of active units, making them more available for major contingencies.
- Fewer forfeited training opportunities for active units.
- New opportunities for National Guard and Reserve units to gain field experience.

This employment would also have some disadvantages:

- Increased cost, largely from the difference between normal pay and active-duty pay. This cost should be added to Reserve and National Guard accounts, not be borne by the active component.
- More active duty away from home, a disadvantage that could be mitigated by preferring volunteers.

National Guard and Reserve personnel might be selected and mobilized in several different ways. By way of illustration, a state National

Guard might designate a particular unit for peacekeeping well in advance of its mobilization. This unit might be one battalion within a National Guard division. Personnel who were unwilling to go on active duty—for example, self-employed persons who would suffer financial loss—could request reassignment to other units. Alternatively, a provisional battalion could be formed from volunteers. For several months prior to the battalion's scheduled mobilization, personnel would train individually to accomplish mission-essential tasks and be individually tested. Upon activation, the unit would train at squad, company, and, finally, at battalion level. Active-duty officers and noncommissioned officers would assist training and assume some positions, possibly including command. The unit would require about a month of post-mobilization training before deploying to its area of operations. It would conduct peacekeeping for six months or possibly a year before returning to the United States. After deployment, it would require another month to reconstitute and allow personnel to take authorized leave.

DEVELOP MODULARITY BELOW THE UNIT LEVEL

The Army and Air Force continue to develop modularity[1] below the unit level.

This option would entail little additional expense and, therefore, few regrets if the level of operations declines. In division- and corps-sized operations, deployment is normally by unit for the obvious reason that units are designed to conduct operations on this scale. But in the smaller deployments thus far characteristic of the period since the Cold War, deployment has been brigade-sized and smaller, causing much fragmentation of units. Flexibility is not the issue: The Army could hardly be more flexible in its task-organization of forces. Indeed, modularity tends to limit flexibility by offering preconceived entities to the planner. But it may be advisable to sacrifice some flexibility in order to gain smoother, more-predictable execution.

---

[1]Modularity implies designing packages below the unit level—for example, earthmovers within an engineer battalion—for separate deployment.

Modularity has least relevance to combat units, because they are already fungible down to the lowest practical levels, e.g., fire teams within an infantry company. It has the greatest relevance to support units that have been increasingly required to perform discrete portions of their overall missions. Implicit in modularity is an understanding of transportation required to deploy, sustainment in the field, and doctrinal statements of the capabilities and limitations of modules.

INCREASE READINESS OF SELECTED ARMY SUPPORT UNITS

The Army increases the readiness of frequently deployed and relatively scarce support units.[2]

This option is moderately sensitive to assumptions about the level of future operations. If the level remains at least constant, this option would be advisable; if the level declines, then the Army would incur additional expense without commensurate gain.

Army units are accorded authorized levels of organization (ALO) in conjunction with anticipated requirements. Support units outside the maneuver divisions are generally accorded lower ALO, as are many units in the Reserve and National Guard. ALO is an authorization, not a guarantee that units will actually attain the authorized level of readiness.[3] Unanticipated resource constraints may cause units to fall short. In addition, some personnel are usually nondeployable—for example, because of illness or temporary duty—and some equipment is usually unserviceable. Therefore, even units at ALO 1 are usually refreshed prior to deployment, if time permits.

---

[2]The Army is already moving in this direction. See Maj. Gen. David L. Grange and Col. Benton H. Borum, "The Readiness Factor: A Prescription for Preparing the Army for All Contemporary Challenges," *Armed Forces Journal International*, April 1997.

[3]Army units report their actual readiness through C-ratings, which reflect percentages of personnel and equipment on hand. C-1 is at least 90 percent; C-2 is at least 80 percent; C-3 is at least 70 percent of personnel and 65 percent of equipment; C-4 is anything less than C-3. Thus, a unit allocated ALO 1 that reported C-1 might have 10 percent shortfalls in personnel and equipment at the time it reported. Support units may be deployed at low levels of readiness. For example, support units deployed during DESERT SHIELD and DESERT STORM had to be only C-3.

During the Cold War, many types of support units, especially those in general support,[4] were required at such infrequent intervals that it was sound policy to conserve resources by keeping them less ready. But since the end of the Cold War, frequent smaller-scale contingencies have upset this calculation. At an increased operational tempo, there are fewer opportunities to conserve resources, because units have to be made ready before they deploy. Moreover, when less-ready units are called up, they usually have to draw personnel and equipment from sister units ("cross-leveling"), causing turbulence and further reducing readiness in the losing units, perhaps to the point that these units can no longer train effectively.[5]

Selection of units would depend on multiple factors, including complementary types (e.g., topographical companies at various echelons), current readiness levels, geographic locations, wartime missions, scheduled changes in Tables of Organization and Equipment/Modified Tables of Organization and Equipment (TOE/MTOE), and relative scarcity across the components. For example, the Army could improve military police support by increasing readiness among the large pool of units in the active component. By contrast, civil affairs resides almost entirely in the reserve component: There is just one civil affairs battalion in the active Army. A decision to increase the readiness of a Reserve or National Guard unit would not necessarily imply that it would deploy abroad: It might be slated to replace an active unit that deployed, a process called "backfilling." Based on recent deployment patterns, the types of support units listed in Figure 5.1 stand out as appropriate candidates.

---

[4]Generally speaking, the Army sought to maintain units in direct support at the same ALO as the units they supported, while maintaining units in general support at lower ALO.

[5]Cross-leveling is a pervasive process that also affects units at higher levels of readiness. Those units are likely to have at least some unreliable equipment and inefficient personnel that they will want to swap before deploying. Charles Barry noted in his review of the draft report that "the impact of cross-leveling on non-deploying or later deploying units is substantial. Often they become 'not-ready' as they swap out personnel and equipment with first deploying units. Eventually you get to units that are much less cohesive and well trained, or even non-deployable. But in an era of smaller peace operations, the stay behind units rarely if ever deploy. Instead, they languish as a continuing source of resources for the deployed force. . . ."

RANDMR951-5.1

Unit Description	Branch	SRC	AC	RC	NG	Total
Combat Engineer Support Company	Engineer	05423L	7	5	16	28
Medium-Girder Bridging Company	Engineer	05463L	2	5	6	13
Assault Bridging Company	Engineer	05493L	5	2	8	15
Firefighting Detachment	Engineer	05510LA	1	12	8	21
Firefighting Company	Engineer	05510LB	7	—	—	7
Topographical Company (Theater)	Engineer	05606L	2	—	1	3
Topographical Company (Corps)	Engineer	05607L	2	—	1	3
HHD Military Police Battalion	Military Police	19476L	10	—	6	16
Military Police Company (Combat Support)	Military Police	19477L	44	—	34	77
Military Police Company (Combat Support)	Military Police	19667L	—	6	13	19
Psychological Ops Company (Strat Dissem)	Psychological Ops	33715A	1	1	—	2
Civil Affairs Battalion (General-Purpose)	Civil Affairs	41735L	1	—	—	1
Civil Affairs Battalion (General-Purpose)	Civil Affairs	41715L	—	21	—	21
Civil Affairs Bn (Foreign Internal Defense)	Civil Affairs	41715L	—	3	—	3
Medical Logistics Battalion (Forward)	Medical	08485L	2	3	1	6
Ammunition Company (General-Support—PLS)	Ordnance	09433L	2	7	—	9
Ammunition Company (Direct Support)	Ordnance	09484L	6	6	—	12
Ammunition Company (General-Support—PLS)	Ordnance	09633L	1	3	—	4
HHD Petroleum Supply Battalion	Quartermaster	10426L	1	7	6	14
Petroleum Supply Company	Quartermaster	10427L	4	22	2	28
Water-Purification Detachment	Quartermaster	10570LC	5	8	11	24
Light–Medium Truck Company	Transportation	55719L	6	2	11	19
Terminal Services Company	Transportation	55827L	4	5	—	9

SOURCE: Component entries were compiled from the Structure and Manpower Allocation System (SAMAS) database, which is current to September 1996, i.e., without regard to transactions planned to occur after that time. The candidates are frequently employed types of units that make important contributions to humanitarian intervention and peace operations. In most cases, a Standard Requirements Code (SRC) to the sixth field, i.e., the series number of the TOE/MTOE, uniquely identifies a type of unit. But in some cases, an alphabetic designator in the seventh field is required for unique designation.

NOTE: AC = active component; Bn = battalion; HHD = Headquarters and Headquarters Detachment; NG = National Guard; Ops = Operations; PLS = Palletized Load System; RC = reserve component; SRC = Standard Requirements Code; Strat Dissem = Strategic Dissemination.

Figure 5.1—Candidate Units for Increased Readiness

ADD SUPPORT UNITS TO THE ACTIVE ARMY

The Army adds some frequently deployed low-density support units to the active component.

This option is highly sensitive to assumptions about the level of operations. If the level remains at least constant, this option would be advisable; if the level declines, then the Army would regret sacrificing other priorities to add support units that found little employment. The Army would especially regret sacrificing combat power if combat units had to be traded for noncombat units within end-strength limitations, i.e., the legislated limit on overall personnel strength.

Following the Vietnam War, the Army deliberately moved support units to the reserve component, for two major reasons:

First, this move allowed the Army to maintain greater combat power than would otherwise have been possible within a constrained budget. The active component kept a high proportion of combat units and enough support units to initiate large-scale operations. The reserve component acquired enough support units to sustain large-scale operations. This division of labor has functioned well, even at the high level of operations experienced in recent years. Individuals and units from the inactive components have performed competently, and no retention problems have yet emerged.

Second, this move implied that the United States would have to mobilize inactive components during war. Mobilization would affect wider circles of the civilian population and presumably compel an administration to seek national support for its policies. While more important in conflicts such as Vietnam, this rationale might also apply to coercive peace operations.

Although no serious problems have yet emerged, it is not clear whether National Guard and Reserve could sustain the current level of operations indefinitely. They were intended to support exceptional operations that would occur infrequently, not nearly continuous operations. If called up too frequently, National Guardsmen and Reservists would eventually begin to resign rather than accept so much disruption of their lives.

Selection of units would be influenced by planning for major theater warfare as well as for the operations described in this report. Units that would be useful early during a large-scale mobilization, either deploying or as backfill, would have priority. In addition, the Army would have to consider whether some types of high-demand units could be kept properly trained at acceptable cost within the active force. Civil affairs, for example, demands skills that are more easily maintained in civilian life than on active military service. Among the types of units that would be candidates are military police battalion headquarters, military police combat support companies, psychological operations battalions, civil affairs battalions, public affairs detachments, and movement-control detachments.

DEVELOP AIR EXPEDITIONARY FORCES FOR CLOSE AIR SUPPORT

The Air Force develops air expeditionary forces optimized to provide the close air support often required during humanitarian intervention and coercive peace operations.

Such air expeditionary forces would be equally useful in other operations of comparable size, including unilateral interventions. Since the Air Force has already developed the Air Expeditionary Force (AEF) concept and has repeatedly deployed such forces, this option would probably entail little additional expense.

AEFs designed for postwar operations in Iraq have been optimized to maintain air superiority and to conduct ground attack without reference to friendly forces. But operations in the former Yugoslavia[6] have required close support of land forces when they were challenged by parties to the conflict or by lawless elements. To provide such support, an AEF would need forward air controllers, both airborne and on the ground, plus a systems-and-munitions mix optimized for the mission. In addition, it would need a command element, probably including an airborne command post, responsive to requests from land forces. Such a force might deploy during the first phase of an operation or stay available for rapid deployment in a crisis.

---

[6]An AEF has not been used in these operations (support to the United Nations Protection Force, JOINT ENDEAVOR, JOINT GUARD) because the area of operations is easily accessible from NATO air bases and because foreign units have composed a large proportion of the force. But under other circumstances, an AEF optimized for close support might be highly appropriate.

MAKE INCREASED USE OF UNMANNED AERIAL VEHICLES

The Air Force promotes development and use of unmanned aerial vehicles (UAVs) to diminish the demand for manned platforms in reconnaissance, electronic warfare, and other missions.

This option is insensitive to assumptions about the level of operations considered in this report. Unmanned platforms are highly desirable in a wide range of situations and will undoubtedly be developed even if the level of these operations declines.

In 1996, the Air Force assumed responsibility for Predator, a medium-altitude (up to 25,000 feet) vehicle with in-flight reprogramming and remote piloting; it transmits real-time video and synthetic aperture radar images. Predator flew over Bosnia-Herzegovina during U.N.–led operations and later to support JOINT ENDEAVOR/JOINT GUARD. In addition, the Air Force is currently developing Global Hawk, a high-altitude (up to 65,000 feet) vehicle, and DarkStar, a high-altitude vehicle with stealth characteristics for use in high-threat environments. Eventually, UAVs will help diminish requirements for manned reconnaissance flights.

SUMMARY OF OPTIONS

Most options are fairly insensitive to the level of operations. If the types of operations described in this report were to decline in frequency and size, most options would remain desirable or at least unobjectionable. (See Figure 5.2.) Two exceptions concern Army units: (1) Increasing the readiness of support units, whether in the active or inactive component, would yield benefit only if operations remained at least at the current level, and (2) adding support units to the active component might be counterproductive if the level declined. Anticipating problems that have not yet fully emerged would leave the Army with resources allocated to support units that might better have been allocated to modernization, among other pressing needs.

Most of the options would not only improve the conduct of humanitarian interventions and coercive peace operations, they would also improve power projection for other purposes, including unilateral interventions like JUST CAUSE and U.S.–led multilateral operations like URGENT FURY.

RANDMR951-5.2

Option	Agencies	Sensitivity to Level of These Operations	Help Improve Power Projection?
Refine command elements for combined joint task forces.	Joint Staff, unified commands, services	Low	Yes
Use a variety of specialized forces to perform search and rescue.	Unified commands, services	Low	—
Expand selective use of civilian contractors.	Unified commands, services	Low	Yes
Organize Army contingency brigades during peacetime.	Army	Low	Yes
Use National Guard and Reserve for noncoercive peacekeeping.	Army	Low	—
Develop modularity below the unit level.	Army, Air Force	Low	Yes
Increase readiness of selected Army support units.	Army	Medium	Yes
Add support units to the active Army.	Army	High	—
Develop Air Expeditionary Forces optimized for close air support.	Air Force, Army	Low	Yes
Make increased use of unmanned surveillance platforms.	Services	Low	Yes

Figure 5.2—Recommendations

Three options stand out as especially helpful in this broader context:

- Further development of command elements for combined joint task forces
- Organization of Army contingency brigades
- Development of air expeditionary forces optimized for close air support.

Contingency brigades and air expeditionary forces are a natural fit and have strong synergistic effects. Together, they would be a powerful, versatile force appropriate for a wide range of contingencies.

Appendix A

OPERATIONS, 1990–1996

This appendix contains a list of operations extracted from the Operations Table in the Force Access database (Table A.1) that were considered in compiling this report. The first column gives either the code name if applicable (e.g., PROJECT HANDCLASP) or a descriptive name (e.g., Multinational Force and Observers). The second column gives the category of operation as described in Appendix B. The third column gives a brief statement of the mission. The fourth column gives start and end dates for the operation so far as known.

Sources for this data include the following:

- *Air Mobility Command Historical Chronologies,* Scott Air Force Base, Ill.: Headquarters, Air Mobility Command, 1997.
- Electronic database prepared by Defense Forecast Incorporated (DFI), 1997.
- U.S. Marine Corps, Headquarters (POC-30), *Marine Corps Operations Since 1776,* Washington, D.C., 1996.
- C. M. Robbins, *Surface Combatant Requirements for Military Operations Other Than War (MOOTW),* Baltimore, Md.: The Johns Hopkins University, 1996.
- George Stewart, Scott M. Fabbri, and Adam B. Siegel, *JTF Operations Since 1983,* Alexandria, Va.: Center for Naval Analyses, 1994.
- Adam B. Siegel, *A Chronology of U.S. Marine Corps Humanitarian Assistance and Peace Operations,* Alexandria, Va.: Center for Naval Analyses, 1994.

- *U.S. Army Deployments Since World II,* The Center of Military History, 1997.
- Command histories prepared by the U.S. Pacific Command and U.S. Air Forces in Europe, on file with the Joint Staff Historical Office, The Pentagon, Washington, D.C.

Table A.1

Operations Table

Operation	Category	Mission	Dates
Multinational Force and Observers	Traditional Peacekeeping	Observe demilitarized zone in the Sinai Peninsula of Egypt and report violations.	25 Apr 82–present
Afghan Refugees	Humanitarian Airlift	Airlift aid to refugees in Afghanistan.	Jan 90–Dec 92
Paraguay Relief	Humanitarian Airlift	Airlift relief supplies following a cyclone.	Feb 90
Ivory Coast Relief	Humanitarian Airlift	Airlift medical supplies and clothing.	Feb 90
Typhoon Offa	Humanitarian Airlift	Airlift supplies to American Samoa following Typhoon Offa.	6–10 Feb 90
Hurricane Hugo	OCONUS HA	Provide relief in Antigua: deliver medical supplies; clear roads; restore electrical power; provide potable water.	18 Apr 90–2 May 90
California Forest Fire 1	CONUS HA	Assist in fighting fires in California and Oregon.	25 Jun 90–1 Aug 90
Philippine Earthquake	OCONUS HA	Assist following earthquake in Luzon: search for survivors; provide relief supplies.	18 Jul 90–30 Jul 90
Tennessee Forest Fire	CONUS HA	Assist in fighting forest fires in Tennessee.	Aug 90
Maritime Interception Operations	Sanctions	Ensure implementation of UNSC Resolution 661 regarding commerce with Iraq.	17 Aug 90–28 Feb 91
Typhoon Mike	OCONUS HA	Provide relief following Typhoon Mike in Philippines.	26 Nov 90–8 Dec 90
Typhoon Owen	Humanitarian Airlift	Airlift relief supplies to Guam.	Dec 90

Table A.1—continued

Operation	Category	Mission	Dates
Nicaragua Assistance	Humanitarian Airlift	Airlift medical supplies to Nicaragua.	Jan 91–May 92
Laos Assistance	Humanitarian Airlift	Airlift medical supplies.	Feb 91
Liberia Assistance 1	Humanitarian Airlift	Airlift supplies to Monrovia following violent coup d'etat.	Feb 91
Sierra Leone Assistance	Humanitarian Airlift	Airlift humanitarian aid.	Feb–Nov 91
Central African Republic Flights	Operational Airlift	Airlift 600 French troops to the Central African Republic to keep order.	26–27 Feb 91
Armenia Earthquake	Humanitarian Airlift	Airlift supplies to Armenia, Turkey, following an earthquake.	Mar 91
PROJECT HANDCLASP	Humanitarian Airlift	Airlift humanitarian aid to Romania.	Mar–Dec 91
Kuwait Oil Fire	Humanitarian Airlift	Airlift firefighting equipment to Kuwait.	Mar–Jun 91
Peru Cholera	Humanitarian Airlift	Airlift medical supplies.	Apr 91
PROVIDE COMFORT I	Humanitarian Intervention	Protect humanitarian assistance to Kurdish population of northern Iraq under UNSC Resolution 688.	6 Apr 91–15 Jul 91
SEA ANGEL	OCONUS HA	Distribute aid after Typhoon Marian struck Bangladesh.	11 May 91–13 Jun 91
Mongolia Flood	Humanitarian Airlift	Airlift medical supplies following a flood.	1 Jun 91–2 Oct 91
Ethiopia Drought	Humanitarian Airlift	Airlift food and medical supplies during drought.	Jun–Sep 91
FIERY VIGIL	OCONUS HA	Evacuate 21,000 U.S. citizens from Clark Air Force Base following eruption of Mount Pinatubo.	8–30 Jun 91
Chad Drought	Humanitarian Airlift	Airlift to N'Djamena during drought and civil conflict	Jul 91

Table A.1—continued

Operation	Category	Mission	Dates
PROVIDE COMFORT II	No-Fly Zone	Enforce no-fly zone north of 36 degrees.	16 Jul 91–present
Albania Relief	Humanitarian Airlift	Airlift food to Tirana.	21 Jul 91–30 Aug 91
China Flood	Humanitarian Airlift	Airlift medical supplies to Shanghai following a flood.	9 Aug 91
Angola Assistance	Humanitarian Airlift	Airlift relief supplies to Angola during civil conflict.	Oct–Nov 91
Ukraine Assistance	Humanitarian Airlift	Airlift medical and other supplies to Kiev.	23–31 Oct 91
North Pole Rescue	OCONUS HA	Rescue 14 Canadian survivors of C-130 crash near North Pole.	Nov 91
Typhoon Yuri	Humanitarian Airlift	Airlift supplies to Guam.	Nov 91
SAFE HARBOR	Migrants	Provide relief to Haitians fleeing Haiti and assist Immigration and Naturalization Service.	13 Nov 91–30 Jun 93
Hurricane Val	Humanitarian Airlift	Airlift supplies to Samoa.	Dec 91
Liberia Assistance 2	Humanitarian Airlift	Airlift assistance during civil conflict.	Dec 91
Typhoon Zelda	Humanitarian Airlift	Airlift relief to Marshall Islands following a typhoon.	Dec 91
PROVIDE HOPE I	OCONUS HA	Deliver foodstuffs and medical supplies to Commonwealth of Independent States (CIS).	10–26 Feb 92
Turkey Earthquake	Humanitarian Airlift	Airlift food, water, clothing, and heavy equipment following earthquake.	Mar–Apr 92
Uzbekistan Oil Fire	Humanitarian Airlift	Airlift firefighting equipment.	Apr 92
El Salvador Assistance	Humanitarian Airlift	Airlift humanitarian aid.	Apr 92

Table A.1—continued

Operation	Category	Mission	Dates
Nicaragua Volcano	Humanitarian Airlift	Airlift supplies following an eruption.	Apr 92
Chicago Flood	CONUS HA	Provide relief after flooding in Illinois.	Apr 92
PROVIDE HOPE II	OCONUS HA	Deliver supplies by air, commercial rail, and commercial ship to CIS.	3 Apr 92–24 Jul 92
HOT ROCK	OCONUS HA	Lift concrete slabs to divert lava flow of Mount Etna, Sicily.	13 Apr 92
GARDEN PLOT (JTF Los Angeles)	MSCA	Support law enforcement authorities in Los Angeles during Rodney King riots.	1–12 May 92
MARITIME MONITOR	Sanctions	Monitor compliance with UNSC Resolutions concerning arms traffic with former Yugoslavia.	1 Jul 92–14 Jun 93
PROVIDE PROMISE	Humanitarian Intervention	Provide humanitarian aid in former Yugoslavia: put up field hospital in Zagreb; airlift aid to Sarajevo; airdrop aid in Muslim-held enclaves in Bosnia.	3 Jul 92–1 Oct 94
SOUTHERN WATCH	No-Fly Zone	Enforce no-fly zone in southern Iraq south of 32 degrees.	1 Aug 92–present
PROVIDE TRANSITION	Peace Accord	Conduct airlift to support repatriation of demobilized soldiers in Angola.	5 Aug 92–8 Oct 92
PROVIDE RELIEF	OCONUS HA	Airlift aid to Somalia and to Somali refugees in Kenya during a famine caused by civil conflict.	15 Aug 92–2 Dec 92

Table A.1—continued

Operation	Category	Mission	Dates
Hurricane Andrew (JTF Andrew)	CONUS HA	Provide relief after Hurricane Andrew struck Florida and Louisiana: assess damage; remove debris; provide electrical power, emergency rations, medical aid, and temporary housing.	25 Aug 92– 15 Oct 92
JTF Marianas	OCONUS HA	Provide relief after Typhoon Omar struck Guam.	28 Aug 92– 19 Sep 92
Chernobyl Airlift	Humanitarian Intervention	Airlift children suffering from the Chernobyl disaster for treatment.	Sep 92
CLEAN SWEEP (JTF Hawaii)	OCONUS HA	Provide relief after Typhoon Iniki struck the Hawaiian Islands: provide food, water, shelter, medical care, electrical power; clear debris.	12 Sep 92– 6 Oct 92
IMPRESSIVE LIFT I, II	Operational Airlift	Airlift Pakistani forces to Somalia during UNOSOM I.	13–29 Sep 92
Armenia Famine	Humanitarian Airlift	Airlift flour to Armenia, Turkey, during famine.	4–11 Nov 92
Bolivia Mudslide	OCONUS HA	Provide relief after catastrophic mudslide in Bolivia.	Dec 92
RESTORE HOPE	Humanitarian Intervention	Ensure uninhibited movement of relief supplies in Somalia during conflict; assist NGOs.	3 Dec 92– 4 May 93
Pakistan Flood	Humanitarian Airlift	Airlift relief to Islamabad following a flood.	6–20 Dec 92
California Flood	CONUS HA	Assist local authorities in rescuing flood victims.	16–18 Jan 93
PROVIDE REFUGE	Migrants	Rescue 525 Chinese nationals attempting to enter U.S. illegally by the ship *Eastwood*.	5 Feb 93– 6 Mar 93

Table A.1—continued

Operation	Category	Mission	Dates
DENY FLIGHT	No-Fly Zone	Enforce no-fly zone in Bosnia-Herzegovina under UNSC Resolution 816; provide CAS to U.N.–led forces; conduct air strikes to protect "safe areas."	12 Apr 93–20 Dec 95
Cambodia Election	Operational Airlift	Airlift troops and equipment to assist U.N. Transitional Authority in Cambodia (UNTAC) in monitoring elections.	1–29 May 93
CONTINUE HOPE	Peace Accord	Support UNOSOM II: provide Quick Reaction Force and logistics support.	5 May 93–31 Mar 94
SHARP GUARD/ MARITIME GUARD	Sanctions	Enforce sanctions imposed by UNSC on the former Yugoslavia.	15 Jun 93–1 Oct 96
ABLE SENTRY	Traditional Peacekeeping	Observe northern borders of Macedonia with Albania and Serbia.	1 Jul 93–present
UNPROFOR Lift	Operational Airlift	Airlift military personnel to Croatia and Bosnia.	5 Jul 93–31 Dec 93
Midwest Flood	CONUS HA	Provide relief and assistance in Iowa, Illinois, and Missouri.	11 Jul 93–1 Aug 93
California Forest Fire 2	CONUS HA	Assist civilian agencies fighting forest fires.	Aug–Sep 93
Guam Earthquake	OCONUS HA	Provide relief following an earthquake: provide electrical power; remove debris.	8–11 Aug 93
Nepal Flood	Humanitarian Airlift	Airlift Bailey bridge from Mildenhall, England, to Kathmandu following a flood.	11–15 Aug 93
Tunisian Fire	OCONUS HA	Assist in fighting forest fires.	22–23 Aug 93
Amtrak Derailment	CONUS HA	Search for victims of derailment near Mobile, Alabama.	22–24 Sep 93

Table A.1—continued

Operation	Category	Mission	Dates
India Earthquake	Humanitarian Airlift	Airlift supplies following an earthquake.	4–5 Oct 93
Nepal Lift	Operational Airlift	Airlift Nepalese troops to Somalia during Second U.N. Operation in Somalia (UNOSOM II).	5–30 Oct 93
SEA SIGNAL/ SUPPORT DEMOCRACY	Sanctions	Enforce sanctions imposed by UNSC on Cedras regime in Haiti; intercept Haitians attempting to enter the U.S. illegally.	18 Oct 93– 19 Sep 94
Los Angeles Earthquake	CONUS HA	Provide relief following Los Angeles earthquake.	Jan 94
Rwandan Relief	OCONUS HA	Provide relief supplies to Rwandan refugees.	11–17 May 94
Rwanda Peace Operation	Operational Airlift	Airlift armored vehicles to support U.N.–led operation.	22–30 Jun 94
OPERATION WILDFIRE	CONUS HA	Fight forest fires and clear debris in Montana and Washington.	Jul–Sep 94
Hurricane Alberto	CONUS HA	Assist civilian agencies providing relief following Hurricane Alberto in Alabama, Georgia, and Florida.	8 Jul 94
SUPPORT HOPE	OCONUS HA	Provide aid to refugees from Rwanda following a massacre of civilian Tutsis and a Tutsi-led invasion of Rwanda.	17 Jul 94–6 Oct 94
ABLE VIGIL	Migrants	Intercept civilians fleeing Cuba and attempting to enter U.S. illegally.	1 Aug 94–16 Sep 94
Hurricane John	OCONUS HA	Evacuate inhabitants of Johnston Atoll.	24–25 Aug 94
SAFE HAVEN/ SAFE PASSAGE	Migrants	Provide camps for Cubans in Panama; after riots, transport Cubans to Guantánamo Bay.	1 Sep 94–20 Feb 95

Table A.1—continued

Operation	Category	Mission	Dates
UPHOLD DEMOCRACY/ MAINTAIN DEMOCRACY	Peace Accord	Restore and support the legitimate government of President Aristide in Haiti.	19 Sep 94–30 Mar 95
DISTANT HAVEN	Migrants	Transport Haitian refugees to Surinam.	Oct 94
PROVIDE HOPE IV	Humanitarian Airlift	Airlift hospital equipment to Kazakhstan.	Oct 94
UNITED SHIELD	Peace Accord	Secure evacuation of UNOSOM II from Somalia.	7 Jan 95–25 Mar 95
SAFE BORDER	Traditional Peacekeeping	Support Military Observer Mission in Ecuador and Peru (MOMEP) in monitoring a cease-fire.	1 Mar 95–9 Jun 96
RESTORE DEMOCRACY	Peace Accord	Participate in United Nations Mission in Haiti (UNMIH) to support the legitimate government.	31 Mar 95–15 Apr 96
Oklahoma City Bombing	CONUS HA	Provide relief following the bombing of a federal building in Oklahoma City.	19 Apr 95–1 May 95
QUICK LIFT	Operational Airlift	Airlift allied Rapid Reaction Force to Croatia.	30 Jun 95–4 Jul 95
PROMPT RETURN	Migrants	Hold, on Wake Island, Chinese nationals intercepted in an attempt to enter U.S. illegally.	26 Jul 95–31 Aug 95
VIGILANT SENTINEL	Peace Accord	Enforce UNSC resolutions by show of force to deter Iraqi aggression.	17 Aug 95–31 Dec 95
DELIBERATE FORCE/ DEADEYE	Air Strike	Conduct air strikes against Bosnian Serb forces threatening Muslim "safe areas" in Bosnia.	29 Aug 95–20 Sep 95

Table A.1—continued

Operation	Category	Mission	Dates
CARIBBEAN EXPRESS	OCONUS HA	Provide relief after Hurricane Marilyn struck Puerto Rico and Virgin Islands.	16 Sep 95–10 Oct 95
Hurricane Opal	CONUS HA	Assist civilian agencies providing relief after Hurricane Opal struck Georgia and Florida.	5 Oct 95
JOINT ENDEAVOR	Peace Accord	Enforce provisions of the Dayton Agreements, especially cease-fire, withdrawal from zone of separation, and cantonment of heavy weapons.	5 Dec 95–20 Dec 96
Northeast Flood 1	CONUS HA	Assist civilian agencies providing relief in Maryland, Pennsylvania, and West Virginia.	Jan 96
Olympic Bombing	CONUS HA	Assist civilian agencies in aftermath of bombing during Summer Olympic Games in Atlanta, Georgia.	Jul 96
Hurricane Bertha	OCONUS HA	Assist civilian agencies in Puerto Rico and the U.S. Virgin Islands.	Jul 96
TWA Crash	CONUS HA	Recover victims and wreckage of TWA Flight 800 that crashed in Atlantic on 17 July 1996.	24 Jul 96–1 Nov 96
California Forest Fire 3	CONUS HA	Assist civilian agencies fighting forest fires.	Aug 96
Hurricane Fran	CONUS HA	Assist civilian agencies providing relief in South Carolina, North Carolina, and Virginia.	Sep 96
Northeast Flood 2	CONUS HA	Assist civilian agencies in New York following a flood.	Nov 96

Table A.1—continued

Operation	Category	Mission	Dates
Gas Explosion	OCONUS HA	Provide airlift and search & rescue following natural-gas explosion in San Juan, Puerto Rico.	Nov 96
JOINT GUARD	Peace Accord	Enforce provisions of Dayton Agreements.	21 Dec 96–present

NOTES: CAS=close air support; CONUS=continental United States; HA=humanitarian assistance; MSCA=military support to civil authorities; NGO=nongovernmental organization; OCONUS=outside the continental United States; UNOSOM=United Nations Operation in Somalia; UNPROFOR=United Nations Protection Force; UNSC=United Nations Security Council.

Appendix B

THE FORCE ACCESS DATABASE

This appendix describes the Force Access database outlined in Chapter Two. A relational database, Force Access combines data on operations with data on units in all four services. When fully developed, it will give a comprehensive historical record of disaster relief, humanitarian assistance, and peace operations from 1990 through 1996, as well as an initial look into the force structure of the services conducting these operations. Force Access employs commercial software and requires little or no training to use.

TECHNICAL SPECIFICATIONS

Hardware

Force Access is designed for use on a stand-alone IBM-compatible personnel computer. For optimal performance, the following hardware is recommended:

- Pentium processor.
- Minimum 12 megabytes (MB) of random-access memory (RAM).
- One 3.5-inch high-density disk drive.
- Video Graphics Adapter or higher-resolution video adapter.
- Pointing device.

Software

Force Access was built on Microsoft ACCESS, a commercial software program designed to construct and manage a relational database, using the Windows 95 operating system. Microsoft ACCESS 7.0 typically requires 32-MB of hard-disk space to install. In its current (December 1996) configuration, Force Access requires approximately 10 MB of RAM.

User Expertise

Force Access requires a basic familiarity with the environment of such Microsoft products as Microsoft Word and EXCEL. Any user who is comfortable in this commonly used environment can employ Force Access effectively, with little or no additional training. Microsoft ACCESS incorporates online help features that answer most questions. For additional help, the user should consult one of the commercially available guides such as Roger Jennings, *Using Microsoft ACCESS,* Indianapolis, Indiana: QUE Corporation, 1995.

ARCHITECTURE

Overview

Microsoft ACCESS is a relational database management system (RDBMS), which means that an underlying system relates all data; no additional code is required. The building blocks are tables organized into rows and columns. Each table stores data on a particular subject—e.g., the Operations Table in Force Access stores data relating to distinct military operations. Within a table, horizontal rows ("records") relate to a single instance of the subject, e.g., the row beginning with the Operational Identifier "MNFO01" in the Operations Table relates to the Multinational Force and Observers operating in the Sinai since 1982. Vertical columns ("fields," or "variables") store discrete elements of data. For a relational database (RDB) to perform properly, each record must contain data related to just one subject. A user fetches information from the database using Microsoft ACCESS queries and forms. The output of a query is returned in the form of a new table.

Basic Tables

There are three kinds of basic tables in Force Access: Unit Tables, Force Tables, and an Operations Table.

Units of the armed services are stored in Unit Tables, i.e., UNIT_ARMY, UNIT_AF, UNIT_NAVY, and UNIT_MC. (By convention, the internal names of tables are written in capital letters.) For example, UNIT_ARMY contains one record for each unit in the active Army, National Guard, and Reserve, including deactivated units and certain units of historical interest. Within Unit Tables, each record is headed by a unique unit identifier (U_ID). The U_ID for Army units is their Unit Identification Code (UIC).

Units of the armed services are related to operations in Force Tables: FORCE_ARMY, FORCE _AF, FORCE_NAVY, and FORCE _MC. Force Tables capture the participation of units in operations by matching U_ID to operational identifiers (O_ID).

Operations that have involved the armed services are depicted in the Operations Table. These operations currently include disaster relief, humanitarian assistance, and peace operations.

Relationship of the Basic Tables

The basic tables in Force Access are related to each other through the intersection of Unit Identifiers (U_ID) and Operation Identifiers (O_ID) in the Force Tables. (See Figure B.1.)

When related tables are joined by common fields, referential integrity can be invoked to enforce the uniqueness of records in the primary table. *Referential integrity* means that all records in a primary table ("parents") must be unique and that records in secondary tables ("children") must relate to parents. When referential integrity is enforced, the system does not permit children that do not relate to parents ("orphans"). For example, U_ID in the Unit Tables and O_ID in the Operations Table are parents and therefore must be unique. There can be one and only one U_ID for each unit recorded in the Unit Tables. There can be one and only one O_ID for each operation recorded in the Operations Table. Further, there can be no orphans

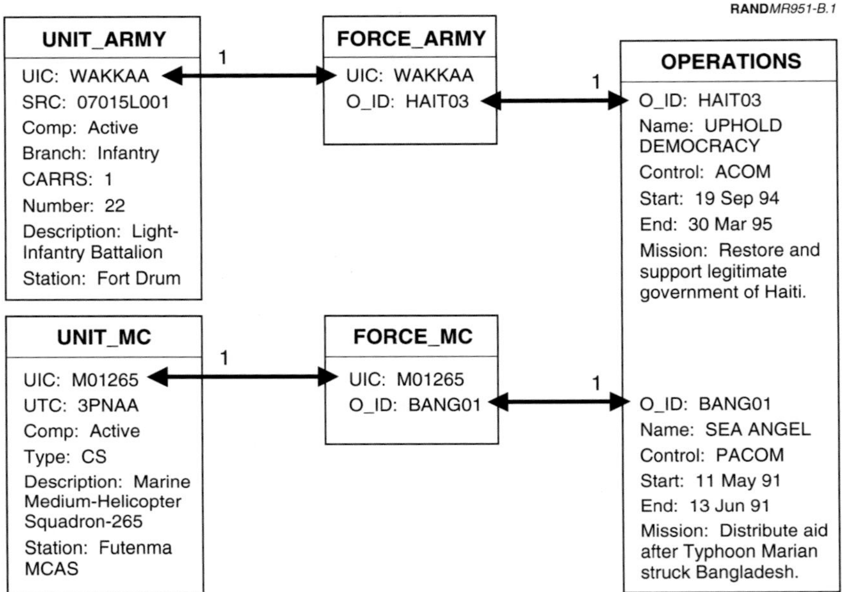

NOTE: ACOM = Atlantic Command; Comp = Component; CARRS = Combat Arms Regimental System; CS = Combat Support; MCAS = Marine Corps Air Station; PACOM = Pacific Command; UIC = Unit Identification Code; UTC = Unit Type Code. The number 1 and the symbol for infinity indicate whether relationships are one-to-many or many-to-one. For example, a given UIC may appear only once in UNIT_ARMY but many times in FORCE_ARMY.

Figure B.1—Intersection in the Force Tables

in secondary tables, e.g., there can be no U_ID in FORCE_ARMY that does not also occur in UNIT_ARMY.

Note also that relationships can be one-to-one, many-to-one, or one-to-many. U_ID in UNIT_ARMY is joined to U_ID in FORCE_ARMY in a one-to-many relationship, depicted in ACCESS by "1" and "∞" on the lines that join tables ("joins"). Thus, there can be more than one occurrence of the same U_ID in FORCE_ARMY to depict the participation of one unit in several operations, or its repeated participation in the same operation at different times. Similarly, O_ID in FORCE_ARMY is joined to O_ID in the Operations Table in a many-to-one relationship. Thus, there can be more than one occurrence of an O_ID in FORCE_ARMY to tag the numerous units that may be involved.

OPERATIONS TABLE

The Operations Table identifies specific operations and briefly describes them. It is linked to all Force Tables through the Operational Identifier (O_ID), and referential integrity is enforced. Currently, the Operations Table contains only those operations subsumed under disaster relief, humanitarian assistance, and peace operations. It does not include other types of operations, such as exercises, noncombatant evacuations, unilateral interventions, or enforcement actions (DESERT SHIELD/DESERT STORM). A user could modify the Operations Table to include additional types of operations or create new tables to reflect them. The Operations Table has the following columns:

OPERATIONAL IDENTIFIER (O_ID)

A 6-character alphanumeric code used to identify a specific operation. The first four characters refer to the geographic area, e.g., "ADRI" refers to the Adriatic Sea, and the last two characters reflect the sequence of entry into the database. Each Operational Identifier is uniquely associated with an Operational Name.

Operational Name (O_Name)

The name, or code name, for a specific operation. Code names are generated by the Joint Staff, unified and specified commands, and major operational commands. For example, "JOINT ENDEAVOR" refers to the military operation to facilitate and enforce implementation of the Dayton Agreements in the former Yugoslavia. In those instances in which we have failed to learn the code name or none was given, the Operational Name is merely descriptive. For example, "Amtrak Derailment" refers to efforts by Marine Corps personnel to locate and rescue victims of an Amtrak derailment in Alabama during September 1993.

Operational Type (O_Type)

The type of operation, according to the following definitions:

Disaster Relief (DR). Operations intended to assist victims of a natural disaster, catastrophic accident, or isolated act of terrorism in a benign environment. Local authorities can maintain or rapidly restore civil order.[1] An example is SEA ANGEL, relief to the population of Bangladesh following Typhoon Marian in April 1991.

Humanitarian Assistance—Noncoercive (AN). Operations intended to relieve suffering caused by conflict, with consent of parties to the conflict. Civil order is seriously disrupted or destroyed by the conflict. The mission or mandate and rules of engagement include self-defense of the force while accomplishing its mission. If conducted under authority of the Security Council, Article VI of the Charter is normally invoked. An example is SUPPORT HOPE, assistance to persons who fled following the massacre of Tutsi civilians and subsequent success of Tutsi rebel forces in Rwanda in June–July 1994.

Humanitarian Assistance—Coercive (AC). Operations intended to relieve suffering caused by conflict, usually with consent of parties to the conflict. Civil order is seriously disrupted or destroyed by the conflict. The mission or mandate and rules of engagement include coercion, if necessary, of persons or parties that attempt to obstruct assistance. If conducted under authority of the Security Council, Article VII of the Charter is normally invoked. An example is RESTORE HOPE, assistance to the citizens of Somalia who in late 1992 were suffering disease and starvation caused by a protracted civil war and exacerbated by drought.

---

[1]The joint definition of *humanitarian assistance* includes disaster relief, noncoercive humanitarian assistance, and coercive humanitarian assistance as defined in this study: "Humanitarian assistance: Programs conducted to relieve or reduce the results of natural or manmade disasters or other endemic conditions such as human pain, disease, hunger or privation that might present a serious threat to life or result in great damage to or loss of property." Joint Chiefs of Staff, *Department of Defense Dictionary of Military and Associated Terms*, Washington, D.C.: U.S. Government Printing Office, Joint Publication 1-02, May 1994. In consultation with the sponsor, we developed study definitions to capture crucial distinctions, especially as concerns the use of force.

Peace Operations—Noncoercive (PN). Operations intended to facilitate a peace process, with consent of the parties to the conflict. If the force is armed, the mission or mandate and rules of engagement typically include only self-defense while accomplishing the mission. If conducted under authority of the Security Council, Article VI of the Charter is normally invoked. An example is the Multinational Force and Observers, which has observed the demilitarized zone in the Sinai Peninsula since 1982. This force is expected to report violations of agreements, but not to coerce the parties (Egypt and Israel). In the event of massive violations, the force would probably withdraw.

Peace Operations—Coercive (PC). Operations intended to facilitate and to enforce, if necessary, a peace process. In all historical cases, the parties have initially given their consent. The mission or mandate and rules of engagement include coercion, if necessary, of any party that attempts to obstruct the peace process. If conducted under authority of the Security Council, Article VI of the Charter is normally invoked. An example is UPHOLD DEMOCRACY, an operation to restore the legitimate government of Haiti in accordance with an agreement concluded by the Carter delegation with the Cedras regime. As events transpired, the Cedras regime maintained consent and agreed to its own demise; had it withdrawn consent, U.S. forces were authorized to compel return of the legitimate government using force as required.

Operational Control (O_Control)

The overall operational command, usually the supported commander-in-chief. Currently, the following entries are allowed: United Nations (UN), U.S. Atlantic Command (USACOM), U.S. European Command (USEUCOM), U.S. Forces Command (USFORSCOM), U.S. Pacific Command (USPACOM), U.S. Southern Command (USSOUTHCOM), U.S. Transportation Command (USTRANSCOM), and Other.

Operational Mission (O_Mission)

A statement of the intended purpose of an operation. Mission statements are contained in, among others, resolutions of the

Security Council, statements by the National Command Authority (NCA), and formulations by the supported commander. There may, of course, be disparity between mission statements and performance.

Operational Start Date (O_SDate)

If a combined or joint task force was created, the official date that it was established. Otherwise, the date that operations actually commenced.

Operational End Date (O_EDate)

If a combined or joint task force was created, the official date that it was disestablished. Otherwise, the date that operations actually ceased.

FORCE ARMY TABLE (FORCE_ARMY)

Associates units, uniquely identified by Unit Identifiers (U_ID), with operations, uniquely identified by Operational Identifiers (O_ID). Referential integrity is enforced for U_ID and O_ID. For Army units, the Unit Identifier is the Unit Identification Code.

Unit Identifier (U_ID)

A 6-character alphanumeric code used to identify a specific unit (see the Unit Army Table).

Operation Identifier (O_ID)

A 6-character alphanumeric code used to identify a specific operation.

Participation Start Date (P_SDate)

Date in day-month-year format, e.g., "1-Jan-96," that a unit began participation in an operation. Participation begins on one of the following:

1. For units in Time-Phased Force Deployment Data (TPFDD), the latest arrival date (LAD). In case of multiple dates for the same unit, the earliest date is used.
2. For units that deployed, but TPFDD is not available, the date determined by other sources, such as a command history.
3. For units that participated, but did not deploy, the start date of the operation.
4. Default is start date of the operation.

Participation End Date (P_EDate)

Date in day-month-year format that a unit ended its participation. Participation ends on one of the following:

1. For deployed units, date of return.
2. For nondeployed units, date that participation ceased.
3. Default is end date of the operation.

Passengers (PAX)

Number of passengers displayed in TPFDD or another documentary source.

Elements Deployed (Elements_Deployed)

A free text field to record information about a unit's participation, e.g., "Company A" of a battalion.

Source ID (Source_ID)

Source of the data entered for a unit in this table, such as a TPFDD or command history.

FORCE AIR FORCE TABLE (FORCE_AF)

Associates units, uniquely identified by Unit Identifiers (U_ID) with operations, uniquely identified by Operational Identifiers (O_ID). Referential integrity is enforced for U_ID and O_ID.

Unit Identifier (U_ID)

A 6-character alphanumeric code used to identify a specific Air Force unit (see UNIT_AF).

Operation Identifier (O_ID)

A 6-character alphanumeric code used to identify a specific operation.

Participation Start Date (P_SDate)

Date in day-month-year format that a unit began participation in an operation. Participation begins on one of the following:

1. For deploying units, date of deployment to a forward operating base.
2. For nondeploying units, date that participation began.
3. Default is start date of the operation.

Participation End Date (P_EDate)

Date in day-month-year format that a unit ended its participation. Participation ends on one of the following:

1. For deployed units, date of return.
2. For nondeployed units, date that participation ceased.
3. Default is end date of the operation.

Aircraft Type (Acft_Type)

A 10-character alphanumeric code that identifies the type of aircraft that participated, e.g., "F-15E," "F-16G."

Aircraft Type Number (Acft_Type_Number)

Number of aircraft of specified type(s) deployed.

Source ID (Source_ID)

Source of the data entered for a unit in this table.

FORCE NAVY TABLE (FORCE_NAVY)

Associates units, uniquely identified by Unit Identifiers (U_ID), with operations, uniquely identified by Operational Identifiers (O_ID). Referential integrity is enforced for U_ID and O_ID.

Unit Identifier (U_ID)

An 8-character alphanumeric code used to identify a specific naval unit (see the Unit Navy Table).

Operation Identifier (O_ID)

A 6-character alphanumeric code used to identify a specific operation.

Participation Start Date (P_SDate)

Date in day-month-year format that a unit began participation in an operation. Participation begins on one of the following:

1. For naval units at sea, date that unit came under operational control.
2. For naval units deploying on land, such as naval mobile construction battalions, date that the unit arrived.
3. Default is start date of the operation.

Participation End Date (P_EDate)

Date in day-month-year format that a unit ended its participation. Participation ends on one of the following:

1. For naval units at sea, date that unit was released from operational control.
2. For naval units deploying on land, date that unit departed.
3. Default is end date of the operation.

Passengers (PAX)

Number of passengers according to TPFDD or manpower according to other documentary source.

Elements Deployed (Elements_Deployed)

A free text field to record information about a unit's participation.

Source ID (Source_ID)

Documentary source of the data entered for a unit in this table.

FORCE MARINE CORPS TABLE (FORCE_MC)

The Force Marine Corps Table associates units, uniquely identified by Unit Identifiers (U_ID), with operations, uniquely identified by Operational Identifiers (O_ID). Referential integrity is enforced for U_ID and O_ID.

Unit Identifier (U_ID)

A 6-character alphanumeric code that identifies a specific unit. (See the Unit Marine Corps Table.)

Operation Identifier (O_ID)

A 6-character alphanumeric code used to identify a specific operation.

Participation Start Date (P_SDate)

Date in day-month-year format that a unit began participation in an operation. Participation begins on one of the following:

1. For units in TPFDD, the latest arrival date (LAD). In case of multiple dates for the same unit, the earliest of those dates.
2. For units that deployed, but for which TPFDD is not available, the date determined by other documentary sources.
3. For units that participated, but did not deploy, the start date of the operation.
4. Default is start date of the operation.

Participation End Date (P_EDate)

Date in day-month-year format that a unit ended its participation. Participation ends on one of the following:

1. For deployed units, date of return.
2. For nondeployed units, date that participation ceased.
3. Default is end date of the operation.

Passengers (PAX)

Number of passengers according to TPFDD or manpower according to other documentary source.

Elements Deployed (Elements_Deployed)

A free text field to record information about a unit's participation, e.g., "Company A" of a battalion.

Source ID (Source_ID)

Documentary source of the data entered for a unit in this table.

UNIT ARMY TABLE (UNIT_ARMY)

Unit data for all currently existing Army units, including deactivated units, was obtained from the Army's Structure and Manpower Allocation System (SAMAS), except as noted. SAMAS contains multiple records for each Unit Identification Code, reflecting previous and planned transactions. We filtered SAMAS, retaining only the data elements noted below for the latest record prior to September 30, 1996. The resulting file displays just force structure; no transaction data, previous or planned, other than deactivation is given. Data for historical units are derived from TPFDD. (See the entry for Current_Force below.)

Unit Identifier (U_ID)

The Unit Identifier for Army units is the Unit Identification Code (UIC), a 6-position alphanumeric code that uniquely identifies a unit organized under a Table of Organization and Equipment (TOE), Modified Table of Organization and Equipment (MTOE), or Table of Distribution and Allowances (TDA). (Broadly speaking, TOE/MTOE define units; TDA define augmentation, such as support organizations in the rear echelons.) For example, WAKKAA uniquely identifies the 1st Battalion, 22nd Infantry (Light). The six positions are used as follows:

Position 1:	Service designator. ("W" indicates an Army unit.)
Positions 2–4:	Parent-unit designators.
Positions 5–6:	Alpha/alpha for parent units; alphanumeric for subunits; numeric/numeric for augmentation units.

Area-Country (Area_Country_Code)

Countries outside the United States are designated by a 2-character code, e.g., "AC" designates Antigua. States within the United States are designated by a 3-character code beginning with the number "1," e.g., "1AL" designates Alabama. Bodies of water are designated by a 3-character code beginning with the number "2," e.g., "2AD" designates the Adriatic Sea. Area_Country_Codes are stored in AREA_COUNTRY.

Service (U_Service_Code)

A 1-character alphabetic code identifies the unit's service affiliation: "A" is given for all units in the Unit Army Table.

Unit Component (U_Component_Code)

A 2-character alphabetic code identifies the component: active (AC), Reserve (RC), and National Guard (NG).

Unit Branch (U_Br_Code)

A 2-character code designates the unit's branch affiliation, e.g., "AV" designates aviation.

Combat Arms Regimental System (CARRS)

The Combat Arms Regimental System assigns a code to combat and to certain combat support units, thus providing an audit trail with an historical regiment. The entry for CARRS is the numerical designation of a unit—for example, "1" for the 1st Battalion.

NUMBER (U_NUMBER)

For TOE/MTOE units, this entry reflects the numerical portion of the parent-unit description—for example, "7" for the 7th Cavalry Regiment. (With few exceptions, the Army has deactivated its regiments, but their designations are retained in CARRS for the purposes of lineage and honors; i.e., the Combat Arms Regimental System continues to carry these units as though they still exist. For example, although the 7th Cavalry has been deactivated, there exist battalions of the 7th Cavalry that trace their lineage to the 7th Cavalry and inherit its honors.) For TDA units, this entry reflects the first four positions of the UIC. For TDA organizations, this entry reflects the numerical designation of the unit being augmented.

Description (U_Descrip)

A shortened title employing abbreviations. For example, "BN LT" describes an infantry battalion (light).

Type of Unit (U_Type)

There are three unit types: TOE/MTOE, TDA, and TDA Augmentation to other units.

Standard Requirements Code (SRC)

The Standard Requirements Code identifies the unit's Table of Organization and Equipment, any variations in TOE, and the level of organization. The positions are used as follows:

Positions 1–2: Branch affiliation, e.g., "01" designates aviation.

Positions 3–5: Organizational elements of the branch or major subdivision. The fifth position indicates the level of organization—e.g., "1" indicates a regiment, brigade, group, or comparable organization.

Position 6: Suffix indicating the TOE series.

Position 7: Last digit of year TOE was published

Positions 8–9: Variations; if standard, the value is "00".

Authorized Strength (Str_Auth)

Manpower reflected in the authorization columns of current or projected authorization documents.

Home Station (U_Home)

The name or abbreviated name of the post, camp, or station where a unit is located.

Current Force (Current_Force)

A true/false field that is checked (true) for all units currently maintained in the active Army, Reserve, or National Guard. It is left blank (false) for units that have been deactivated or removed from the Army's rolls. We added to the Unit Army Table a small number of units that participated in recent operations but have since been deactivated or removed.

Source of Data (Source_ID)

A numeric code that identifies the primary source of unit information.

UNIT AIR FORCE TABLE (UNIT_AF)

Unit Identification Code (U_ID)

A 6-character alphabetic code beginning with "f" that uniquely identifies an Air Force unit.

Country (Area_Country_Code)

A 3-character alphanumeric code used to identify the country of the unit (see AREA_COUNTRY).

Service (U_Service_Code)

A 1-character alphabetic code that identifies the unit's service affiliation: "F" for all units in the Unit Air Force Table.

Major Command (Maj_Com)

A 3-character code that reflects the unit's major command, e.g., Pacific Air Forces ("PAF"). Air Force Reserve is coded "AFR," and Air Force National Guard is coded "ANG" or "NGS."

Unit Component (U_Component_Code)

A 2-character alphabetic code that identifies the component. Components include active (AC), Reserve (RC), and National Guard (NG).

Unit Branch (U_Br_Code)

A 3-character alphabetic code for the type of activity, e.g., "rec" indicates recruiting.

Echelon

A 2-character alphabetic code for the level of organization, e.g., "sq" indicates a squadron.

Number (U_Number)

Four or fewer characters that designate a unit—e.g., "343" designates the 343rd Recruiting Squadron.

Unit Description (U_Descrip)

A shortened title employing abbreviations.

Primary Aerospace Vehicle Authorization (PAA)

The number of assets (aircraft, missiles, etc.) authorized to fulfill a unit's primary mission.

Required Strength (Str_Reqd)

Full authorized strength for the unit.

Home Station (U_Home)

Concatenation of a 4-digit installation code and an installation name.

Current Force (Current_Force)

A true/false field that is checked (true) for all units currently maintained in the active Air Force, Reserve, or Air National Guard. It is left blank (false) for units that have been deactivated or deleted entirely.

Source of Data (Source_ID)

A numeric code that identifies the primary source of unit information.

UNIT NAVY TABLE (UNIT_NAVY)

Unit data for all Navy ships currently in commission, which are obtained from the "List of U.S. Navy Ships" maintained by the Navy Public Affairs Library. This database gives basic information on Navy ships, including names, types, and home ports.

Unit Identifier (U_ID)

A 6-character alphanumeric code that gives ship type and hull number, e.g., "CVN-69."

Country (Area_Country_Code)

A 3-character alphanumeric code that identifies the country. (See AREA_COUNTRY.)

Service (U_Service_Code)

A 1-character alphabetic code that identifies the unit's service affiliation: "N" for all units in the Unit Navy Table.

Unit Component (U_Component_Code)

A 2-character code that identifies the component. All entries in the Unit Navy Table are in the active component ("AC").

Unit Description (U_Description)

Free text field that contains the full name of a ship, e.g., "USS Dwight D. Eisenhower."

Ship Type (U_Ship_Type)

An alphabetic code that identifies the ship type, e.g., "CVN" identifies a nuclear-powered aircraft carrier.

Home Station (U_Home_Station)

A free text field that gives a ship's home port, e.g., "Norfolk, VA."

Current Force (Current_Force)

A true/false field that is checked (true) for ships currently in commission. It is left blank (false) for ships no longer in commission.

Source of Data (Source_ID)

A numeric code that identifies the primary source of the data.

UNIT MARINE CORPS TABLE (UNIT_MC)

Unit data for all Marine Corps units, which are obtained from the Status of Resources and Training System (SORTS) database. SORTS, which is used primarily to report readiness, includes all Marine Corps units down to battalion and separate company/detachment level.

Unit Identifier (U_ID)

A 6-character alphanumeric Unit Identification Code, beginning with "M," that uniquely identifies a unit in the active or reserve Marine Corps.

Country (Area_Country_Code)

A 3-character alphanumeric code used to identify the country of the unit. (See AREA_COUNTRY.)

Service (U_Service_Code)

A 1-character alphabetic code that identifies the unit's service affiliation: "M" for all units in the Unit Marine Corps Table.

Unit Component (U_Component_Code)

A 2-character code that identifies the component. Components include active ("AC") and reserve ("RC").

Function

A 2- to 3-character alphabetic code as follows: combat ("CBT"), combat support ("CS"), combat service support ("CSS").

Unit Level Code (ULC)

A 1- to 3-character code reflecting the level of organization.

Unit Type Code (UTC)

A 5-character alphanumeric code indicating the type of unit—e.g., "OGTAA" indicates a Marine infantry battalion.

Unit Description (U_Descrip)

Free text containing the numerical designation and the full title of a unit, e.g., "2nd Marine Aircraft Wing."

Home Station (U_Home)

A 30-character alphanumeric entry that shows the location at which the unit is garrisoned.

State (U_State)

State in which the unit's home station is located.

Current Force (Current_Force)

A true/false field that is checked (true) for all units currently maintained in the active Marine Corps or Reserve. It is left blank (false) for units that have been deactivated or deleted entirely.

Source of Data (Source_ID)

A numeric code that identifies the primary source of unit information.

TABLES THAT DESCRIBE ASPECTS OF OPERATIONS

Operations have aspects that are not captured by enumeration of the participating units. Force Access currently includes two tables that are joined to the Operations Table and describe aspects of air support.

Operation Air Combat (OPN_AIRCBT)

This table contains data on air combat that occurred during an operation. The fields include aircraft types, numbers of missions flown by type (electronic combat, air-defense suppression, combat air patrol, ground attack, etc.), and sources for these data.

Operation Airlift (OPN_AIRLIFT)

This table contains data on airlifts conducted during an operation. The fields include aircraft types, missions flown, number of passengers, tons of cargo, and sources for these data.

TABLES THAT ALLOW ONE-TO-MANY RELATIONSHIPS

Additional tables are provided to allow one-to-many relationships between key variables while preserving referential integrity.

Operation Area (OPN_AREA)

This table joins the Operations Table to the Area-Country Table through the key field Operation Identifier (O_ID). It allows the geographic aspect of an operation to be described by linking an operation to Area-Country Codes (Area_Country_Code) in a one-to-many relationship, i.e., one operation can encompass many countries and areas.

Operation Data Source (OPN_DATA_SRC)

This table joins the Operations Table to the Data Source Table through the key field Operation Identifier (O_ID). It allows the data sources to be depicted by linking an operation to Source Identifiers (Source_ID) in a one-to-many relationship, i.e., one operation can have many data sources.

TABLES THAT EXPAND VARIABLES

For efficiency and to save screen space, the tables, forms, and queries use codes. Some of these codes are expanded online by pull-down lists that are generated by tables.

Operation Type (OPN_TYPE)

This table expands the codes for five types of operations currently contained in Force Access: Disaster Relief (DR), Humanitarian Assistance—Noncoercive (AN), Humanitarian Assistance—Coercive (AC), Peace Operations—Noncoercive (PN), and Peace Operations—Coercive (PC). It generates a pull-down list, available by clicking.

Operational Control (OPN_CONTROL)

This table is joined to the Operations Table by the key variable Operational Control (O_Control). Operational Control reflects the command entity responsible for the conduct of operations, e.g., the North Atlantic Treaty Organization ("NATO"), United States European Command ("USEUCOM"). This table generates a pull-down list, accessed by clicking.

Area-Country Table (AREA_COUNTRY)

This table contains information on the areas and countries in which operations occurred. It is joined to the Operations Table through the Operation Area Table described above.

Unit Branch (UNT_BRANCH)

This table is joined to the Unit Army Table. It expands codes for branches that identify broad military skill groups, e.g., armor (AR), field artillery (FA), and infantry (IN).

Unit Level (UNT_LEVEL)

This table is joined to the Unit Marine Table. It expands codes that identify organizational levels in the Marine Corps, e.g., division, battalion, and company.

Unit Ship Type (UNT_SHIP_TYPE)

This table is joined to the Unit Navy Table. It expands codes that identify ship types, e.g., ammunition ship (AE).

UTILITY OF FORCE ACCESS

Force Access provides a powerful combination of operational history and force structure within an easily used relational database. Fully developed, it will offer an unprecedented look into past operations and a useful tool to explore the implications for force mix and force structure. Even in its current state of development, it provides a useful overview of past efforts, especially those at the high end (Somalia, Haiti, Bosnia), and ways to compare this level of effort with available forces. When Force Access is demonstrated, observers often wonder why such a database has not been available. Had one been developed in 1990 and maintained to the current time, U.S. planners would have been spared much uncertainty about the requirements for post–Cold War operations. We believe that Force Access, properly developed and maintained, will be an invaluable asset for planners in the Office of the Secretary of Defense, the Joint Staff, and the services beyond the life of this project.

INHERENT LIMITATIONS

Force Access is inherently limited by its architecture, which is built on the participation of units in operations. Unit participation reveals much about capabilities, but it becomes problematic when units are created *ad hoc*, extensively tailored, or rotated incrementally.

For example, during smaller operations, the Air Force seldom deploys or employs complete units. Because airlift is fungible, it often makes little sense to assign a mission exclusively to particular squadrons or wings. To support deployment to undeveloped air bases, the Air Force typically deploys certain wing assets, but not entire support units. When an operation extends over months, the Air Force usually rotates aircraft through the operation, often a few aircraft at a time, rather than entire squadrons. Therefore, a list of participating Air Force units may not be very useful or may require extensive interpretation.

The Army also tailors its forces flexibly, especially the units providing combat support and combat service support. It deploys units at less than full strength and attaches personnel to units as required by the mission. Moreover, the Army may employ units outside the primary mission of those units—for example, using light infantry in a policing role. Marine forces are usually tailored before they put to sea, for example, in the Marine Expeditionary Unit (MEU), which is built around a Marine infantry battalion. But their degree of participation may be difficult to measure—as for example, during SEA ANGEL, when most Marines remained embarked and rotated through the operation on land.

INCOMPLETE DATA

The data in Force Access are incomplete in two respects: (1) changes in unit composition over time and (2) deployment data, including start and end dates.

As to changes in unit composition over time, the Unit Tables in Force Access reflect only current force structure (plus some records for historical Army units). But units change their composition over time. They may be issued more-modern weapons and equipment. They may be reorganized according to revised authorization documents, e.g., MTOE for Army units. As a result, it can be misleading to assume that units in historical operations resemble their current counterparts. This problem is negligible for current operations but becomes more serious further back in time. It could be largely overcome by providing a complete set of Unit Tables for each year covered by the database; however, it is doubtful whether such an extensive effort would yield commensurate gain.

As to incomplete deployment data, Force Access in its current version can only reveal level of effort in the best-understood operations, e.g., Hurricane Andrew, Somalia, Haiti, and the opening phase of operations in Bosnia. Incomplete or unreliable arrival and departure dates mean that Force Access reflects little about the development of a given operation over time. In addition, it is more complete for Army and Marine forces than for Air Force and Navy forces.

Their professional historians notwithstanding, the services seem to have remarkably little interest in preserving records of their deployments. The Army, for example, cannot produce coherent or comprehensive records of deployment during its major post–Cold War operations. To generate the data currently in Force Access, we correlated data on Army deployments from three sources: TPFDD, command histories, and briefings. TPFDD-derived data are highly specific and detailed, but still fragmentary. TPFDD concerns the arrival of forces in the area of operations and gives no information about attachments. Moreover, TPFDD may not always be executed as planned. The better command histories include comprehensive troop lists, but these lists do not include dates of arrival or departure and convey little information about partial deployments. Command briefings give a good sense of the overall force structure and provide a cross-reference, but add little detail.

Appendix C

STRESSES ON ARMY UNITS

This appendix contains the output of a spreadsheet analyzing stress on Army units as described in Chapter Four (see Table C.1).

The first heading has two columns: "Description" and "SRC." "Description" includes the unit's name plus data to establish a unique type, e.g., "105 T" indicates that a field artillery battalion is equipped with towed 105mm howitzers. "SRC" (Standard Requirements Code) identifies branch and Table of Organization and Equipment (TOE) for the type of unit.

The second and third headings refer to the underlying scenario. "No Overlap" implies that operations are accomplished sequentially; "Simultaneous" implies that operations begin at the same time. Each of these headings has three columns: "I/%AC," 1–VI/%AC," and "I–VI/%Total." The first of these columns expresses the requirement over the first six months as a percentage of the active Army's inventory. The second column expresses the requirement over three years, i.e., six increments of six months each, as a percentage of the active Army's inventory. The third column expresses the requirement over three years as a percentage of the total Army inventory (active, Reserve, and National Guard).

Table C.1
Stresses on Army Units

Unit Description	SRC	No Overlap			Simultaneous		
		I % AC	I–VI % AC	I–VI % Tot	I % AC	I–VI % AC	I–VI %Tot
Engineer Battalion (Light Division)	05155L0001E0	50	100	67	50	100	67
HHC Engineer Brigade (Heavy)	05332L000100	0	33	20	17	17	20
Engineer Battalion (Heavy Division)	05335L000100	2	27	15	14	27	15
HHC Engineer Group (EAC)	05412L100100	10	10	1	10	10	1
Combat Engineer Support Company	05423L000100	17	83	19	50	83	19
Medium-Girder Bridging Company	05463L100100	0	150	23	100	50	23
Assault Bridging Company	05493L100100	0	50	14	25	25	14
Firefighting Detachment	05510LA00100	22	100	32	44	78	32
Diving Detachment—Light Weight	05530LC00100	17	17	14	17	17	14
Topographical Company	05607L000100	8	58	38	33	33	38
HHB Division Artillery (Light)	06102L000100	15	15	10	15	15	10
Field Artillery Battalion (105mm Towed)	06125L000100	6	12	8	12	6	8
HHB Division Artillery (Heavy)	06302L000100	0	24	10	12	12	10
Target Acquisition Battery	06303L000100	0	71	36	43	57	36
Field Artillery Battalion (155mm SP)	06365L100100	2	27	12	14	27	12
Target Acquisition Det (Corps)	06413L000100	17	17	14	17	17	14
Infantry Battalion (Light)	07015L000100	11	26	17	11	26	17
Ranger Battalion (Airborne)	07085L000100	0	20	20	0	20	20
Infantry Battalion (Mechanized)	07245L000100	4	26	9	13	39	9
Reconnaissance Battalion (Light Division)	17185L000100	50	50	33	50	50	33
Tank Battalion (M1A1)	17375L000100	4	19	7	11	11	7
Special Forces Battalion (Airborne)	31805L000100	13	33	24	20	37	24
Air-Defense Battalion (Heavy)	44175L100100	0	24	9	12	12	9
HHB Air-Defense Brigade	44412L000100	0	30	24	15	15	24
HHC Division (Light)	77004L000100	30	45	30	30	45	30
HHC Brigade (Light)	77042L000100	25	65	37	25	65	37
HHC Division (Heavy)	87004L200100	0	24	10	12	12	10
HHC Brigade (Heavy)	87042L100100	0	27	12	13	29	12
HHC Division Aviation Brigade (Lt Div)	01102A000100	50	65	43	50	65	43
Attack-Helicopter Battalion (Lt Div)	01175L000100	50	100	100	50	100	100
Medium-Helicopter Battalion	01245A000100	0	31	28	9	23	28
HHC Division Aviation Brigade (Hvy Div)	01302L000100	0	33	17	17	27	17

RAND*MR951-T-C.1a*

Table C.1—continued

Unit Description	SRC	No Overlap			Simultaneous		
		I % AC	I–VI % AC	I–VI % Tot	I % AC	I–VI % AC	I–VI %Tot
General Support Helicopter Battalion	01305A000100	40	140	78	60	170	78
Attack-Helicopter Battalion (Lt Div)	01385A200100	8	25	11	17	17	11
Combat Aviation Battalion (Spec Ops)	01855A000100	0	260	260	100	160	260
Aviation Maintenance Company	01945A300100	10	36	11	16	45	11
Chemical Company (Hvy Div)	03157L200100	0	24	5	12	12	5
Chemical Company (Lt Div)	03467L000100	8	15	12	8	15	12
Signal Battalion Divisional MSE	11065L000100	3	6	4	3	6	4
Signal Battalion Area MSE	11435L000100	0	13	8	0	28	8
Signal Battalion (Corps Support)	11445L100100	0	160	32	60	100	32
Signal Company Tactical Satellite	11603L200100	10	87	87	43	53	87
Signal Brigade (Theater)	11612L000100	6	38	21	12	32	21
Signal Company Tropo (Light)	11667L000100	10	20	7	10	20	7
Signal Company Tropo (Heavy)	11668L000100	100	200	40	100	200	40
MP Company Inf Div (Light)	19323L000100	50	50	33	50	50	33
MP Company Inf Div (Heavy)	19333L000100	0	20	8	20	0	8
HHC Military Police Brigade	19472L000100	0	33	20	0	33	20
HHD Military Police Battalion	19476L000100	3	48	29	14	59	29
MP Company (Combat Support)	19477L000100	3	35	19	13	48	19
MP Company (Guard, Combat Support)	19677L000100	0	75	5	0	200	5
Military Police Detachment	19683L000100	15	130	29	65	80	29
Military History Detachment	20017L000100	100	600	35	300	400	35
HHC Psychological Operations Battalion	33706L000100	20	127	38	53	93	38
Psych Ops Company Strat Dissem	33715A000100	0	100	50	100	0	50
Military Intelligence Bn Inf Div (Light)	34355A100100	10	20	15	10	20	15
Military Intelligence Bn Inf Div (Heavy)	34395A0001E0	0	24	7	12	12	7
HHD MI Brigade (Heavy Corps)	34402L000300	8	45	45	23	30	45
MI Battalion (Aerial Exploitation)	34415L100100	0	40	40	20	20	40
Military Intelligence Company (EPW)	34567AA00300	0	100	100	0	300	100
MI Battalion (EAC)	34665L000000	30	60	60	30	60	60
HQ Civil Affairs Brigade	41702L000200	—	—	10	—	—	10
Civil Affairs Battalion (General-Purpose)	41735L000100	30	260	13	130	160	13
Public Affairs Detachment Mobile	45413L000100	0	800	20	300	900	20

RAND*MR951-T-C.1b*

Table C.1—continued

Unit Description	SRC	No Overlap			Simultaneous		
		I % AC	I–VI % AC	I–VI % Tot	I % AC	I–VI % AC	I–VI %Tot
Public Affairs Team	45500LA00100	33	100	55	50	125	55
Public Affairs Det Broadcasting	45607L000100	—	—	33	—	—	33
Surgical Detachment	08407L100100	33	83	45	50	67	45
Veterinarian Detachment	08419L000100	8	23	14	8	23	14
HHD Medical Brigade (Corps)	08422L100100	0	60	12	30	30	12
Air Ambulance Company	08447L200100	13	50	15	25	63	15
Ground Ambulance Company	08449L000100	25	50	11	25	50	11
Medical Battalion (Area Support)	08455L000100	50	100	29	50	100	29
Dental Services Company	08478L000100	6	32	5	6	32	5
Medical Logistics Battalion (Forward)	08485L000100	50	200	80	100	150	80
Preventive Medicine Det (Sanitation)	08498L000100	50	113	35	50	113	35
Preventive Medicine Det (Entomology)	08499L000100	20	100	38	20	100	38
Combat Stress Control Detachment	08567LA00100	17	67	27	33	50	27
Hospital Combat Support	08705L000100	14	50	13	29	64	13
Hospital Mobile Surgical	08863L000100	0	50	9	0	50	9
Missile Maintenance Company	09428L000100	0	20	15	10	10	15
Ammunition Company PLS GS	09433L000100	0	100	22	50	50	22
Ammunition Company PLS DS	09484L000100	5	10	4	5	10	4
Ordnance Team (EOD)	09527LB00100	9	35	31	13	39	31
Ammunition Company (General-Support)	09633L000100	100	100	25	100	100	25
HHC Petroleum Battalion (Terminal Ops)	10416L000100	50	200	133	100	150	133
HHD Petroleum Supply Battalion	10426L000100	30	260	20	130	160	20
Petroleum Supply Company	10427L000100	15	80	11	40	55	11
Water-Purification Detachment	10570LC00100	80	240	50	120	240	50
Replacement Company	12407L000100	0	25	3	0	25	3
HHD Personnel Services Battalion	12425L100100	8	45	45	23	30	45
Personnel Detachment	12427L000100	2	5	2	3	5	2
Postal Company	12447L000100	33	73	17	53	93	17
Personal Services Company	12467L100100	—	—	33	—	—	33
Finance Group	14412L000100	15	30	12	15	30	12
Finance Detachment	14423L000100	0	14	6	5	14	6
HHD Finance Battalion	14426L100100	2	10	5	5	7	5

RAND*MR951-T-C.1c*

Table C.1—continued

Unit Description	SRC	No Overlap			Simultaneous		
		I % AC	I–VI % AC	I–VI % Tot	I % AC	I–VI % AC	I–VI %Tot
Chaplain Team Support (DS)	16500LB00100	—	—	75	—	—	75
Maintenance Company (Non-Divisional)	43209L000100	1	6	2	3	6	2
Detachment (Cargo Documentation)	55560LA00100	7	29	15	7	29	15
Detachment (Contracts)	55560LC00100	17	33	10	17	33	10
Det (Automatic Cargo Documentation)	55560LD00100	33	67	33	33	67	33
Heavy Crane Detachment	55560LE00100	33	67	67	33	67	67
Movement-Control Detachment	55580LH00100	100	750	37	300	800	37
Movement-Control Center COSCOM	55604L000100	33	43	33	33	43	33
HHC Composite Group	55622L000100	30	240	48	130	140	48
HHD Motor Transport Battalion	55716L000100	20	40	10	20	40	10
Light–Medium Truck Company	55719L200100	26	83	50	26	83	50
Medium Truck Company (EAC)	55727L100100	0	40	7	10	30	7
Medium Truck Company (Corps)	55728L100100	11	67	25	11	58	17
Terminal Services Company	55827L000300	25	33	14	25	33	14
Forward Support Battalion (Heavy)	63005L100100	2	27	14	14	27	14
Forward Support Battalion (Light)	63215L000100	20	80	50	40	120	50
Division Support Command (Light)	63222L000100	15	30	20	15	30	20
Main Support Battalion (Light)	63225L000100	17	18	16	17	18	16
HQ Support Group	63422L000100	9	11	11	9	22	0
HHC Corps Support Command	63431L000100	0	67	40	33	33	40
COSCOM Movement-Control Center	63433L000100	40	35	33	40	60	8
Aviation Support Battalion (Heavy)	63885A200100	0	40	40	20	20	40
Special Operations Support Battalion	63905L000100	60	90	90	60	90	90

NOTE: AC = Active Army; Bn = Battalion; COSCOM = Corps Support Command; Det = Detachment; DS = Direct Support; EAC = Echelons Above Corps; EOD = Explosive Ordnance Disposal; EPW = Enemy Prisoners of War; GS = General Support; HHB = Headquarters and Headquarters Battery; HHC = Headquarters and Headquarters Company; HQ = Headquarters; Hvy Div = Heavy Division; Inf Div = Infantry Division; Lt Div = Light Division; MI = Military Intelligence; MP = Military Police; MSE = Multiple Subscriber Equipment; Ops = Operations; PLS = Palletized Load System; Psych Ops = Psychological Operations; SP = Self-Propelled; Spec Ops = Special Operations; Strat Dissem = Strategic Dissemination; Tropo = Tropospheric.

RAND*MR951-T-C.1d*

Appendix D

ARMY CONTINGENCY BRIGADE

In this appendix, we explore the common characteristics of Army forces during recent high-end operations and describe a notional Army contingency brigade.

COMMON CHARACTERISTICS

Army forces for smaller-scale contingencies are organized uniquely for each operation; however, in three important cases (Somalia, Haiti, Bosnia) they had the following common characteristics:

- Maneuver brigades were central.
- Both heavy and light maneuver forces were included.
- Aviation played a large role, including attack helicopters in Somalia and Bosnia.
- Integral sustainment was provided for protracted operations.

Maneuver Brigades Were Central

The Army is fundamentally organized to conduct division- and corps-sized operations. But none of the contingency operations conducted since 1989, except the Persian Gulf War, required deployment of even one Army division. Instead, these contingencies required task forces built around one or two maneuver brigades; the command element was drawn from divisional headquarters, supplemented from corps assets. Therefore, it would be useful to develop maneuver brigades that are prepared for rapid deployment and autonomous operations.

Heavy and Light Maneuver Forces Were Included

The Army's maneuver battalions are composed of one branch (armor, mechanized infantry, light infantry, air assault, airborne) and are normally organized into either heavy (armor, mechanized infantry) or light (light infantry, air assault, airborne) brigades. But smaller-scale contingencies have usually demanded a mix of heavy *and* light forces. RESTORE HOPE (Somalia) required motorized infantry (brigade from 10th Mountain Division supplied with wheeled vehicles) supported by a few heavy units. The subsequent operation, CONTINUE HOPE, was reinforced by battalions of armor and mechanized infantry. UPHOLD DEMOCRACY (Haiti) required light and motorized infantry, with very limited armor support in the initial phase. JOINT ENDEAVOR (Bosnia) initially required two heavy brigades, each containing one armor battalion and one mechanized infantry battalion, plus military police performing, to some extent, the role of motorized infantry. Judging from this experience, the Army can expect that future contingencies will require similar mixes, not either light or heavy forces alone.

Army Aviation Played a Large Role

Army aviation has played a large and often critical role in smaller-scale contingencies. It was in constant demand for reconnaissance, mobility, and logistics support. In addition, it was frequently required to support air assault operations and to deliver fire support. The most spectacular use of Army aviation occurred during the hunt for Farah Aideed in Somalia during CONTINUE HOPE, but it was just as essential in Haiti and Bosnia. In Bosnia, for example, AH-64 attack helicopters provided near-real-time surveillance of critical terrain and impressive shows of force when elements of the Bosnian Serbs attempted to deny freedom of access to U.S. forces. In future contingencies, Army aviation is almost certain to be in high demand.

Integral Sustainment Was Provided for Protracted Operations

The Army has played a central role in these contingencies because of its capability to project land combat power over large areas of operations and to sustain this power for months or years. To generate this capability, the Army deploys substantial numbers of support units and, in addition, contracts for civilian logistics support. These support elements are simply the price for sustained land combat power, a price that is particularly high for an all-volunteer force drawn from an affluent society. Although the support elements vary from one operation to another, they follow a fairly regular and predictable pattern, which is closely tied to the selection of combat forces.

NOTIONAL STRUCTURE FOR A CONTINGENCY BRIGADE

An Army contingency brigade having the characteristics described above might have a structure such as that in Table D.1.

Table D.1
Notional Army Contingency Brigade

Unit Description	SRC	TOE	%	Personnel
Engineer Battalion (Heavy Division)	05335L000100	436	1	436
Combat Engineer Support Company	05423L000100	178	1	178
Assault Bridging Company	05493L100100	181	1	181
Topographical Company	05607L000100	116	0.1	12
Target Acquisition Battery	06303L000100	79	1	79
Field Artillery Battalion (155mm SP)	06365L100100	641	1	641
Target Acquisition Det (Corps)	06413L000100	39	1	39
Infantry Battalion (Light)	07015L000100	568	1	568
Tank Battalion (M1A1)	17375L000100	633	1	633
Special Forces Battalion (Airborne)	31805L000100	380	0.1	38
Avenger Battalion	44435L000100	377	0.3	113
HHC Corps	52401L300100	324	0.2	65
HHC Division (Heavy)	87004L200100	224	0.2	45
HHC Brigade (Heavy)	87042L100100	80	1	80
Medium Helicopter Battalion	01245A000100	795	1	795
HHC Div Aviation Brigade (Hvy Div)	01302L000100	80	1	80
General-Support Helicopter Battalion	01305A000100	330	1	330
Attack Helicopter Battalion	01385A200100	301	1	301
Aviation Maintenance Company	01945A300100	508	1	508
Chemical Company (Hvy Div)	03157L200100	167	0.3	50
Signal Battalion Area MSE	11435L000100	692	0.3	208
Signal Company TACSAT	11603L200100	103	1	103
Signal Brigade (Theater)	11612L000100	94	0.3	28
Signal Company Tropo (Light)	11667L000100	104	1	104
MP Company Inf Div (Light)	19323L000100	83	1	83
MP Company Inf Div (Heavy)	19333L000100	153	1	153
HHD Military Police Battalion	19476L000100	63	0.5	32
MP Company (Combat Support)	19477L000100	177	1	177
Military History Detachment	20017L000100	3	1	3
MI Battalion Inf Div (Heavy)	34395A0001E0	390	0.3	117
HHD MI Brigade (Heavy Corps)	34402L000300	53	0.3	16
MI Battalion (Aerial Explotation)	34415L100100	353	0.3	106
MI Company (EPW)	34567AA00300	51	0.3	15
Civil Affairs Bn (General-Purpose)	41735L000100	208	0.1	21

RAND*MR951-T-D.1a*

Table D.1—continued

Unit Description	SRC	TOE	%	Personnel
Public Affairs Team	45500LA00100	5	1	5
Medical Battalion (Area Support)	08455L000100	343	1	343
Preventive Medicine Det (Sanitation)	08498L000100	11	1	11
Preventive Medicine Det (Entomology)	08499L000100	11	1	11
Combat Stress Control Detachment	08567LA00100	23	0.5	12
Ammunition Company PLS DS	09484L000100	174	1	174
Ordnance Team (EOD)	09527LB00100	23	1	23
Petroleum Supply Company	10427L000100	197	1	197
Water-Purification Detachment	10570LC00100	15	2	30
Postal Company	12447L000100	52	0.3	16
Personal Services Company	12467L100100	57	0.3	17
Finance Group	14412L000100	66	0.1	7
Finance Detachment	14423L000100	19	1	19
Chaplain Team Support	16500LB00100	2	1	2
Detachment (Contract Supervision)	55560LC00100	12	1	12
Movement-Control Detachment	55580LH00100	4	1	4
HHC Composite Group	55622L000100	98	1	98
Light–Medium Truck Company	55719L200100	108	1	108
Medium Truck Company (EAC)	55727L100100	175	1	175
Forward Support Battalion (Heavy)	63005L100100	436	1	436
Aviation Support Battalion (Heavy)	63885A200100	531	0.5	266
Total Personnel:				8301

SOURCE: Component entries were compiled from the Structure and Manpower Allocation System (SAMAS) database, which is current to September 1996, i.e., without regard to transactions planned to occur after that time. The candidates are frequently employed types of units that make important contributions to humanitarian intervention and peace operations. In most cases, a Standard Requirements Code (SRC) to the sixth field, i.e., the series number of the TOE/MTOE, uniquely identifies a type of unit. But in some cases, an alphabetic designator in the seventh field is required for unique designation.

NOTE: Bn = Battalion; Det = Detachment; Div = Division; DS = Direct Support; EAC = Echelons Above Corps; EOD = Explosive Ordnance Disposal; EPW = Enemy Prisoners of War; HHC = Headquarters and Headquarters Company; HHD = Headquarters and Headquarters Detachment; Hvy = Heavy; Inf Div = Infantry Division; MI = Military Intelligence; MP = Military Police; MSE = Multiple Subscriber Equipment; PLS DS = Palletized Load System Direct Support; SP = Self-Propelled; TACSAT = Tactical Communications Satellite; TOE = Table of Organization and Equipment; Tropo = Tropospheric. Total personnel is calculated by multiplying personnel authorized according to TOE (third column) by a percentage (fourth column), yielding 8301; the Strength column sums to 8304 because of rounding. Of course, an actual brigade would be structured in numbers of people, not percentages of TOE.

RAND*MR951-T-D.1b*

REFERENCES

Anderson, Gary W., Lt. Col. (USMC), *Operation Sea Angel: A Retrospective on the 1991 Humanitarian Relief Operation in Bangladesh*, Newport, R.I.: Naval War College, Strategy and Campaign Department, Report 1-92, 1992.

Barry, Charles L., "Review of Draft RAND Study: Assessing Requirements for Peacekeeping, Humanitarian Assistance, and Disaster Relief, August 1997," unpublished.

Bigelow, James H., Thomas J. Blaschke, James Chiesa, and Adele R. Palmer, "Initial Processing of Army Personnel Force (M-FORCE) Data: A Guide for Force Structure Costing System Data Managers," unpublished RAND research.

Brooks, Billy G., *Historical Analysis of U.S. Military Responses, 1975–1995*, San Diego, Calif.: Science Applications International Corporation (SAIC), October 1, 1996.

Brown, Ronald J., *Humanitarian Operations in Northern Iraq, 1991, with Marines in Operation Provide Comfort*, Washington, D.C.: Headquarters, U.S. Marine Corps, History and Museum Division, 1995.

Campbell, James L., *After Action Report: Task Force Mountain Warrior, Operation Restore Hope/Continue Hope, Somalia, 10 April–7 August 1993*, Fort Drum, N.Y.: Headquarters, 1st Brigade, 10th Mountain Division (Light Infantry), 1994.

Center of Military History, "U.S. Army Deployments Since World War II," spreadsheets prepared by the Staff Support Branch, Washington, D.C., forwarded by letter dated August 21, 1996.

Chapman, Suzanne, "Keeping Pilots in the Cockpits," *Air Force Magazine*, July 1997, pp. 66–69.

Clarke, Walter S., *The Comprehensive Campaign Plan (CCP)*, Washington, D.C.: The Congressional Hunger Center, 1997.

Davis, Lois M., et al., *Army Medical Support for Operations Other Than War*, Santa Monica, Calif.: RAND, MR-773-A, 1996.

Doyle, Francis M., Karen J. Lewis, and Leslie A. Williams, "Named Military Operations from January 1989 to December 1993," Fort Monroe, Va.: Training and Doctrine (TRADOC) Library and Information Network (TRALINET), Headquarters, TRADOC, April 1994.

Gallay, David R., and Charles L. Horne III, *LOGCAP Support in Operation Joint Endeavor: A Review and Analysis*, McLean, Va.: Logistics Management Institute, LG612LN1, 1996.

Gertz, Bill, "Miracle in the Desert," *Air Force Magazine*, January 1997, pp. 60–64.

Grange, Maj. Gen. David L., and Col. Benton H. Borum, "The Readiness Factor: A Prescription for Preparing the Army for All Contemporary Challenges," *Armed Forces Journal International*, April 1997.

Grier, Peter, "Readiness at the Edge," *Air Force Magazine*, June 1997, pp. 58–62.

Grier, Peter, "Stressed Systems," *Air Force Magazine*, July 1997, pp. 52–55.

Grimmett, Richard F., *Instances of Use of United States Armed Forces Abroad, 1798–1995*, Washington, D.C.: Congressional Research Service, 96-119F, The Library of Congress, February 1996.

Hines, Jay E., *USCENTCOM in Somalia: Operations Provide Relief and Restore Hope*, MacDill AFB, Fla.: United States Central Command History Office, November 1994.

Institute for Foreign Policy Analysis, Fletcher School of Law and Diplomacy, Tufts University, *Strategy, Force Structure, and Defense Planning for the Twenty-First Century*, Washington, D.C., May 1997.

Joint Chiefs of Staff, *Department of Defense Dictionary of Military and Associated Terms*, Washington, D.C.: U.S. Government Printing Office, Joint Publication 1-02, May 1994.

Kassing, David, *Transporting the Army for Operation Restore Hope*, Santa Monica, Calif.: RAND, MR-384-A, 1994.

Kirby, Sheila Nataraj, "Selected Reservists in 1992: Attitudes, Perceptions of Unit Readiness, and Potential Problems If Mobilized," unpublished RAND research.

Kugler, Richard L., "Toward a New U.S. Overseas Presence: USAF's Emerging Role in the Joint Team," unpublished RAND research.

Leland, John W., *Operation Provide Hope*, Office of History, Air Mobility Command, Scott AFB, Ill., July 1993.

Lippiatt, Thomas F., et al., *Mobilization and Train-Up Times for Army Reserve Component Support Units*, Santa Monica, Calif.: RAND, MR-125-A, 1992.

McCarthy, Paul A., *Operation Sea Angel: A Case Study*, Santa Monica, Calif.: RAND, MR-374-A, 1994.

Moore, Nancy Y., et al., *Materiel Distribution: Improving Support to Army Operations in Peace and War*, Santa Monica, Calif.: RAND, MR-642-A, 1997.

Reed, Pamela L., J. Matthew Vaccaro, and William J. Durch, *Handbook on United Nations Peace Operations*, Washington, D.C.: The Henry L. Stimson Center, Handbook No. 3, April 1995.

Robbins, C. M., "Surface Combatant Requirements for Military Operations Other Than War (MOOTW)," Baltimore, Md.: The Johns Hopkins University, Naval Warfare Analysis Department, NWA-96-001, 1996.

Roos, John G., "The 'Outsourcing' Boom," *Armed Forces Journal International*, October 1996, pp. 18–20.

Schank, John F., et al., *Identifying and Supporting Joint Duty Assignments: Executive Summary,* Santa Monica, Calif.: RAND, MR-622-JS, 1996.

Schrader, John Y., *The Army's Role in Domestic Disaster Support: An Assessment of Policy Choices,* Santa Monica, Calif.: RAND, MR-303-A, 1993.

Stewart, George, Scott M. Fabbri, and Adam B. Siegel, *JTF Operations Since 1983,* Alexandria, Va.: Center for Naval Analyses, CRM 94-42, July 1994.

Siegel, Adam B., *A Chronology of U.S. Marine Corps Humanitarian Assistance and Peace Operations,* Alexandria, Va.: Center for Naval Analyses, CIM 334, September 1994.

Siegel, Adam B., *Requirements for Humanitarian Assistance and Peace Operations: Insights From Seven Case Studies,* Center for Naval Analyses, CRM 94-74, Alexandria, Va.: March 1995.

Siegel, Adam B., *The Use of Naval Forces in the Post War Era: U.S. Navy and U.S. Marine Corps Crisis Response Activity, 1946–1990,* Alexandria, Va.: Center for Naval Analyses, CRM 90-246, February 1991.

Smith, Charles R., *Angels from the Sea: Relief Operations in Bangladesh, 1991*, Washington, D.C.: Headquarters, U.S. Marine Corps, History and Museums Division, 1995.

Sortor, Ronald E., *Army Active/Reserve Mix, Force Planning for Major Regional Contingencies,* Santa Monica, Calif.: RAND, MR-545-A, 1995.

Sortor, Ronald E., "Army Forces for OOTW," unpublished RAND research.

Sortor, Ronald E., and Gavin Pearson, "Planning Army Force Structures for OOTW: A Status Report," unpublished RAND research.

Taw, Jennifer Morrison, David Persselin, and Maren Leed, *Meeting Peace Operations Requirements While Maintaining MRC Readiness*, Santa Monica, Calif: RAND, MR-921-A, 1998.

Tirpak, John A., "The Expeditionary Air Force Takes Shape," *Air Force Magazine,* June 1997, pp. 28–33.

U.S. Air Force, Headquarters, Air Mobility Command, *1997 Air Mobility Master Plan (1997-AMMP),* Scott AFB, Ill., 1996.

U.S. Department of the Army, TRADOC Analysis Center, *Reconstitution of Army Combat Service Support Units Engaged in Operations Other Than War,* Fort Leavenworth, Kan.: Technical Report TRAC-TR-0995, February 1996.

U.S. Department of the Army, The Center of Military History, *U.S. Army Deployments Since World War II,* Washington, D.C., 1996.

U.S. Department of the Army, TRADOC Analysis Center, *Reconstitution of Army Combat Service Support Units Engaged in Operations Other Than War,* Technical Report TRAC-TR-0995, February 1996.

U.S. Department of the Army, "U.S. Army Major Deployments Since 1989," Washington, D.C.: Army Staff, Directorate of Operations, Readiness and Mobilization, Force Readiness Branch, 1996.

U.S. Department of the Air Force, *45 Years of Global Reach and Power, The United States Air Force and National Security: 1947–1992,* Washington, D.C., 1992.

U.S. General Accounting Office, *Contingency Operations: Opportunities to Improve the Logistics Civil Augmentation Program,* Washington, D.C.: GAO/NSIAD-97-63, February 1997.

U.S. Marine Corps, Headquarters (POC-30), *Marine Corps Operations Since 1776,* Washington, D.C., 1996.

Vick, Alan, David T. Orletsky, Abram N. Shulsky, and John Stillion, *Preparing the U.S. Air Force for Military Operations Other Than War,* Santa Monica, Calif.: RAND, MR-842-AF, 1997.